J. H Tucker

A manual of sugar analysis

including the applications in general of analytical methods to the sugar industry

J. H Tucker

A manual of sugar analysis
including the applications in general of analytical methods to the sugar industry

ISBN/EAN: 9783337414320

Printed in Europe, USA, Canada, Australia, Japan

Cover: Foto ©berggeist007 / pixelio.de

More available books at **www.hansebooks.com**

A

MANUAL

OF

SUGAR ANALYSIS

INCLUDING

THE APPLICATIONS IN GENERAL

OF

ANALYTICAL METHODS TO THE SUGAR INDUSTRY.

WITH AN INTRODUCTION

ON THE

CHEMISTRY OF CANE-SUGAR, DEXTROSE, LEVULOSE, AND MILK-SUGAR.

BY

J. H. TUCKER, PH.D.

SECOND EDITION.

NEW YORK:
D. VAN NOSTRAND, PUBLISHER,
23 MURRAY STREET AND 27 WARREN STREET.
1883.

PREFACE.

NOTWITHSTANDING the amount and variety of analytical work required for the various interests connected with sugar, there exists no book in English that treats of this branch of analysis, and only a few scattered and incomplete dictionary articles. The main dependence of the chemist must be on German and French sources, in which languages treatises on sugar analysis are numerous.

I have accordingly attempted, with as much success as may be, to supply this deficiency in the special literature of analytical chemistry, believing that there is now, and still more in the near future there will be, a need of such a work in English-speaking countries.

An introduction on the chemistry of the more important sugars is given, on account of its intimate connection with the subject; the matter is brought strictly up to the time of publication, and in some important respects—as, for example, in relation to inversion, and the melassigenic action of salts on cane-sugar—I believe to be more full than can be found elsewhere.

The formulas and atomic weights used are according to the new system, and the temperatures are Centigrade.

I desire to render my acknowledgments to Prof. C. F. Chandler, of Columbia College, New York, for access to his fine private library of technical chemistry; to Dr. A. Behr, of Jersey City, for the loan of books, and other favors; and to W. Baker, Esq., Librarian of the School of Mines, Columbia College, New York, for uniform courtesy and help.

THE AUTHOR.

NEW YORK, July 1881.

CONTENTS.

CHAPTER I.

THE CHEMISTRY OF THE SUGARS AS A CLASS.

CHAPTER II.

CANE-SUGAR OR SACCHAROSE.

CHAPTER III.

DEXTROSE, LEVULOSE, AND INVERT-SUGAR.

CHAPTER IV.

LACTOSE OR MILK-SUGAR.

CHAPTER V.

DETERMINATION OF SPECIFIC GRAVITY.

CHAPTER VI.

DETERMINATION OF CANE-SUGAR—OPTICAL METHODS.

CHAPTER VII.

DETERMINATION OF CANE-SUGAR—CHEMICAL METHODS.

CHAPTER VIII.

DETERMINATION OF DEXTROSE AND INVERT-SUGAR.

CHAPTER IX.

ANALYSIS OF RAW SUGAR.

CHAPTER X.

ANALYSIS OF MOLASSES AND SYRUPS.

CHAPTER XI.

ANALYSIS OF THE CANE AND CANE-JUICE.

CHAPTER XII.

ANALYSIS OF THE BEET AND BEET-JUICE.

CHAPTER XIII.

ANALYSIS OF WASTE PRODUCTS.

CHAPTER XIV.

ANALYSIS OF COMMERCIAL GLUCOSE OR STARCH-SUGAR.

CHAPTER XV.

ESTIMATION OF MILK-SUGAR.

CHAPTER XVI.

ESTIMATION OF DEXTROSE IN DIABETIC URINE.

CHAPTER XVII.

THE CHEMISTRY OF ANIMAL CHARCOAL.

CHAPTER XVIII.

THE ANALYSIS OF ANIMAL CHARCOAL.

APPENDIX.

CHAPTER I.

THE CHEMISTRY OF THE SUGARS AS A CLASS.

THE term sugar is applied to a group of bodies resembling each other by a number of striking properties; these properties—partly chemical and partly physical—are as follows: (1) the sweet taste; (2) the ability to undergo the process of fermentation; (3) the identity or similarity in chemical composition or relations; (4) the power that aqueous solutions have of rotating the plane of polarized light; (5) the general resemblance in physical and chemical characteristics, such as their ready solubility in water, insolubility in absolute alcohol and ether, facility of crystallization in well-defined forms, the similarity of their products of oxidation, and their ability to reduce the oxide of copper in alkaline solution, either directly or after conversion into some other sugar by fermentation or the action of dilute acids.

The sweet taste is very distinctive of sugars, and is possessed in a greater or less degree by nearly all of them, from cane-sugar, which is the type of sweet substances, to some of the rarer saccharoids, which have this property in a very low degree or not at all. The sugars are not the only bodies, however, possessing a sweet taste, as glycol and glycerin are sweet, as well as some metallic salts, notably the acetates of lead; the two former are, however, allied to the sugars, as they are polyatomic alcohols. The yttria salts and some silver compounds are also said to have a sugary flavor. The relative sweetening power of cane-sugar to dextrose has been generally placed as two to one; Parmentier questions this, and gives the following quantities of the two sugars as having an identical sweetening effect:

10 pts. of cane-sugar to 40 pts. of water.
12 " dextrose to 40 " "

It has been asserted that levulose is sweeter than cane-sugar, and this seems to be confirmed by the fact that invert-sugar is sweeter than cane-sugar, H. Morton placing the excess at ten per cent.

The sugars are mostly of vegetable origin, though a few, as inosite and dextrose, are found in animals; they exist in a great variety of plants distributed over every part of the globe.

CHEMICAL CONSTITUTION.

All bodies known as sugars are composed of carbon, oxygen, and hydrogen; in the true sugars the hydrogen and oxygen are present in the proportions that form water. For example, dextrose $C_6H_{12}O_6$, has the hydrogen and oxygen present in the exact proportion to make six molecules of water, and the formula may be written thus: $C_6(H_2O)^6$; the compound, from this point of view, may be considered as a hydrate of carbon. Bodies thus constituted are called *carbohydrates*, and under this name are included many important substances not classed as sugars, though sugar may be derived from many of them by the action of diastase or acids. The most important of these are: starch,

$nC_6H_{10}O_5$; cellulose, $nC_6H_{10}O_5$; gum, $C_{12}H_{22}O_{11}$; and dextrin, $C_6H_{10}O_5$.

Most sugars belong to the class of the *hexatomic alcohols* and the corresponding ethers. An alcohol is a compound in which hydrogen, in a saturated hydrocarbon, is replaced by one or more atoms of the univàlent radical hydroxyl HO; thus, propenyl alcohol or glycerin $(C_3H_5)'''(HO)_3$, is derived from the hydrocarbon *propane* C_3H_8 by substituting three atoms of hydroxyl for the same number of hydrogen, the result being a triatomic alcohol —that is, one containing three atoms of hydroxyl in the place of an equal number of replaceable hydrogen. So with higher replacements; mannite $C_6H_{14}O_6$, may be considered as derived from the saturated hydrocarbon C_6H_{14} by replacing six atoms of hydrogen with an equal number of hydroxyl atoms, and the formula may be written $(C_6H_8)^{vi}(HO)^6$, which represents a hexatomic alcohol.

Mannite and dulcite are important representatives of the sugars having the composition of alcohols.

Sugars of the formula $C_6H_{12}O_6$, or the *glucoses*, have two atoms of hydrogen less than the saturated alcohols, and are classed as *aldehydes* of these alcohols; this classification is justified by the fact that dextrose, when acted upon by nascent hydrogen, is converted into mannite, just as acetic aldehyde is changed into ethylic alcohol by the same agent.

Sugars of the composition $C_{12}H_{22}O_{11}$, such as cane-sugar, are so constituted that one molecule is equivalent to two molecules of the glucoses, minus one molecule of water—as, $C_{12}H_{22}O_{11} = (2C_6H_{12}O_6 - OH_2)$; these are called *diglucosic alcohols.* The carbohydrates starch, cellulose, and a few others, having the formula $C_6H_{10}O_5$, or multiples of it, may be regarded as the *oxygen ethers* or *anhydrides* of the glucoses or the diglucosic alcohols, inasmuch as they differ from them by one molecule of water.

The most important of the sugars may be arranged, according to their chemical relations, as follows:

I. SATURATED ALCOHOLS.

Triatomic.	*Pentatomic.*	*Hexatomic.*
Dambonite.	$(C_6H_7)^v\ (OH)^5$	$(C_6H_8)^{vi}(OH)^6$
$(C_4H_6)'''(OH)^3$.	Quercite.	Mannite.
Derived from butylene,	Pinite.	Dulcite.
C_4H_8.	*Derived from the hydrocarbon* C_6H_7.	Isodulcite.
		Rhamnegite.
		Derived from the hydrocarbon C_6H_8.

II. ALDEHYDES OF THE HEXATOMIC ALCOHOLS (GLUCOSES).

$$C_6H_{14}O_6 - H_2 = C_6H_{12}O_6.$$

Dextrose.	Mannitose.	Eucalyn.
Levulose.	Dulcitose.	Inosite.
Galactose.		Dambose.

III. DIGLUCOSIC ALCOHOLS.

[*Related to the glucoses by* $C_{12}H_{22}O_{11} = (2C_6H_{12}O_6 - H_2O)$].

Saccharose.	Melitose.
Parasaccharose.	Mycose. }
Lactose.	Trehalose. }
Melezitose.	Synanthrose.
	Maltose.

The above grouping is based on the modern theories of organic chemistry, which may be useful in this application, as they have undoubtedly been in many others. A more convenient, and possibly an equally scientific classification may be made, based on the relations of constitutional identity or similarity in regard to their empirical formulas, and general chemical and physical properties. The true sugars or carbohydrates have the characters pre-eminently saccharine, while the saccharoids differ much from them in atomic constitution and many other properties. The carbohydrates are classed as **fermentable** and **non-fermentable.** Class I. is again divided into A, *Glucoses*, and B, *Sucroses*, the former being capable of fermenting directly without previous conversion into any other body.

CARBOHYDRATES.						SACCHAROIDS.					
CLASS I. FERMENTABLE. *Either directly or after inversion.*				CLASS II. NON-FERMENTABLE.		CONTAINING EXCESS OF HYDROGEN.					
A. *Glucoses.*		B. *Sucroses.*				I.		II.		III.	
$C_6H_{12}O_6$.	Rot. power.	$C_{12}H_{22}O_{11}$.		$C_6H_{12}O_6$.		$C_6H_{14}O_6$	Rot. power.	$C_6H_{12}O_5$			
Dextrose,	(a) j 56°	Saccharose,	73.8°	Eucalyn, (nearly)	50°	Mannite,	(slight)	Quercite,	33.5°	Borneite, $C_7H_{14}O_6$.	
Levulose,	106° at 14° C.	Lactose,	58.4°	Sorbose,	46.9°	Dulcite,	0°	Pinite,	58.6°	Matezite, $C_{10}H_{20}O_9$.	79°
Galactose,	83°	Melezitose,	94.1°	Nucite,	0°	Isodulcite,	7.6°			Dambonite, $C_4H_8O_3$.	0°
Mannitose,	0°	Melitose,	102°	Arabinose,	116°	Rhamnegite,	26°			Erythromannite, $C_4H_{10}O_4$.	0°
Inosite,	0°	Mycose,	173.2°	Borneose,	0°	Sorbite,	0°			Raffinose, $C_6H_{14}O_7$.	117.3°
		Trehalose,	220°	Dambose, $C_3H_6O_3$	0°					Scyllite.	
		Synanthrose,	0°	Metezose, $C_9H_{18}O_9$	6°						
		Maltose,	150°								

Formation in Plants.—Cane-sugar is probably derived from the starch existing in the plant, which is converted by the action of diastase, or a similar ferment, into the soluble form, or dextrin, and then into sugar by the fixation of the elements of water. According to Payen, all immature parts of the sugar-cane contain starch, while at maturity there is not a trace of it.

M. A. Richard, in his *Précis de Botanique*, gives the following account of the conversion of the amylaceous matter into sugar: "Starch has the same chemical composition as cellulose—carbon and the elements of water. It is found largely in all the organs of the plant, where it accumulates to serve for nutrition; but, like cellulose, starch is insoluble in water. In order to render it assimilable it must be made soluble, and this is effected by a peculiar body, *diastase*, discovered by MM. Payen and Persoz, which exists or is formed under certain circumstances in all organs containing starch. Diastase possesses the peculiar property of converting starch into a saccharoid and soluble matter, dextrin, which is dissolved and carried by the juice to all parts of the plant. Now, the dextrin, combining with one equivalent more of water, is changed into cane-sugar. The latter may be modified in its turn, combining with additional water and producing dextrose, or grape-sugar."

As a result of numerous experiments made by M. Biot* upon the conditions of the formation and change of sugar in various plants, including the sycamore, maple, birch, walnut, and wheat, the following conclusions were arrived at: 1. The existence, at a certain age of the plant, of grape or

* *Compt. Rend.*

invert sugar alone. 2. The simultaneous presence of cane and invert, or grape-sugar. 3. The existence of different sugars in different organs of the same plant. 4. The natural and normal transformation of different sugars into each other, and even into dextrin. 5. The converse of the above—that starch and dextrin are converted into cane-sugar.

Berthelot and Buignet,* in experiments on the orange, have shown the remarkable fact that cane-sugar forms in the presence of free citric acid, which appears to be not only without invertive action, but actually prevents the formation of invert-sugar.

C. T. Jackson† considers, in the case of the sugar millet, that cane is derived from invert-sugar at the period of maturity. To this Icery,‡ from the result of his examination of the growing sugar-cane, assents, but claims an important rôle for the influence of light in the transformation.

Synthetical Studies of the Sugars.—It has long been a favorite idea with chemists to produce the sugars, and especially cane-sugar, by artificial means; but up to the present there has been but small success in this department of research. Honig and Rosenfeld (*Ber. Chem. Gesell.*, x. 871) attempted to produce the alkali and halogen combinations of dextrose. Pohl§ states that sugar may be produced from *assamar* (see page 42) when the aqueous solution is allowed to remain a long time at rest. Assamar, according to Reichenbach, has the formula $C_{24}H_{26}O_{13}$, whence

---

* *Compt. Rend.*, li. 1094.

† *Ibid.*, xlvi. 55.

‡ *Ann. Chim. Phys.*, [4] v. 350.

§ *Jour. Pk. Chem.*, lxxxii. 148.

$$C_{24}H_{38}O_{13} + 9H_2O = 2C_{12}H_{22}O_{11}.$$

As the nature of assamar, however, is entirely unknown, the equation has no theoretical value. Löwig * obtained a fermentable syrup from oxalic ether, and Renard † has found, among the products of the action of electrolytic oxygen on glycerin and dilute sulphuric acid, a body $C_6H_{12}O_6$ which reduces alkaline solution of oxide of copper, and on oxidation gives oxalic acid.

By the action of alkalies on invert-sugar Peligot ‡ recognized a substance among the products which he called *saccharin*. This body has the composition of cane-sugar, $C_{12}H_{22}O_{11}$, and crystallizes in large right-rhombic prisms; the taste is not sweet; soluble in cold water; not fermentable with yeast; largely volatile; does not reduce alkaline solution of oxide of copper; nitric acid oxidizes it to oxalic acid; specific rotatory power [a] D = 93° 5′.§

Optical Rotatory Power.—Aqueous solutions of most sugars have the property of rotating the plane of polarized light either to the right or left. The degree of rotation varies with different sugars, and with the majority the direction of the rotation is to the right and is uninfluenced to any great extent by the temperature. Levulose rotates to the left, and the temperature exercises great influence. A freshly-prepared solution of crystallized dextrose, and some other sugars, possesses a rotation of double the ordinary, but when it is left to itself for some hours,

---

* *Jour. Pk. Chem.*, lxxxiii. 129.

† *Compt. Rend.*, lxxxii. 562.

‡ *Ibid.*, lxxxix. 918 ; xc. 1141.

§ Later investigations of Scheibler (*Neue-Zeits*, v. 261) lead him to assign the formula $C_6H_{10}O_5$; see also Kolli and Vachovic (*Ibid.*, v. 170) on the synthesis of sugars.

or heated, the specific rotatory power is reduced to the normal degree, where it remains constant. This property of sugars is called *Birotation* (see page 79).

RELATION OF SUGARS TO THE PHENOMENA OF FERMENTATION.

The ability to decompose under the influence of a nitrogenous exciting agent or ferment is perhaps the most distinctive characteristic of these bodies. By placing a solution of sugar under suitable conditions the evolution of carbonic acid or hydrogen, and the formation of alcohol or lactic acid, is positive proof of the presence of sugar or saccharoidal matter. As far as known there are no exceptions to this.

The sugars are capable of undergoing various kinds of fermentation, the difference consisting in the nature of the ferment, the temperature, the concentration of the fluid in which the action takes place, and the products formed.

Mucous Fermentation (Gmelin's *Handbook Cav. Soc.*, xv. 280).—This fermentation takes place under the influence of a peculiar mucous ferment which is composed of sporules of .0012 to .0014 mm. in diameter, and, when introduced into cane-sugar solutions containing albumen, causes the sugar to be resolved into mannite, gum, and carbonic acid. 100 pts. of cane-sugar yield, on the average,

59.01 pts. mannite,
45.50 " gum,

corresponding to the equation:

$$25C_{12}H_{22}O_{11} + 13H_2O = 12C_{12}H_{20}O_{10} + 24C_6H_{14}O_6 + 12CO_2$$

Under conditions not accurately known hydrogen is also

evolved. When a greater proportion of gum is formed than that given above, the sporules are larger and are probably a distinct ferment (Pasteur*). Mucous fermentation requires access of air, and likewise the presence of nitrogenous matter; neither acid nor alcohol is produced (Hochstetter). The fermentation is prevented by sulphuric acid, hydrochloric acid, and alum. Fresh beet-juice, on exposure to the air, becomes gummy, and is found to contain mannite, gum, tartaric acid, and uncrystallizable sugar.

According to Plagne,† the juice of the sugar-cane contains a white, non-azotized substance, which becomes brown and moist in contact with the air, is soft and difficult to dry, soluble in water, insoluble in alcohol and ether, and is precipitated from watery solutions by oxide of lead, mercurous salts, and alcohol. It converts cane-sugar into a substance intermediate between starch and gluten, which forms quickly and somewhat abundantly in syrups, rendering them viscid, ductile, and uncrystallizable. If, therefore, the juice, after being treated with lime, is left to stand forty-eight hours, a jelly is produced from which alcohol throws down a soft white precipitate which dries to a nacreous mass, dissolving but sparingly in hot and cold water, even when moist, but swells up again to a transparent jelly, which, treated with nitric acid, yields only oxalic acid. It is not colored by iodine or converted into glucose by acids, and does not give off ammonia when submitted to dry distillation.‡

Lactous or Butyrous Fermentation.—Various su-

---

* *Bull. Soc. Chim*, 1861, 30. † *Journ. Pharm.*, xxvi. 248.

‡ See *Phil. Mag.*, 1846, 28; Scheibler (*Zeit. f. Rubenz*, xxiv. 309).

gars and dextrin, when subjected to the action of particular ferments, are converted into lactic acid, the change consisting in the resolution of the molecule, preceded in some cases by the assumption of the elements of water—as,

$$(1) \quad \underset{\text{Dextrose.}}{C_6H_{12}O_6} = \underset{\text{Lactic acid.}}{2C_3H_6O_3}.$$

$$(2) \quad \underset{\text{Lactose.}}{C_{12}H_{22}O_{11}} + H_2O = \underset{\text{Lactic acid.}}{4C_3H_6O_3}.$$

The lactous fermentation requires a temperature of 20° to 40° C., the presence of water, and certain ferments—viz., albuminous substances in a peculiar state of decomposition, such as caseine, gluten, and animal membranes. The action depends upon a ferment contained in or formed by the above substances, and, according to Blondeau, is the vegetable growth *Penicilium glaucum.*

When the lactous fermentation is set up in a suitable solution, it is because certain bodies present in the air develop the ferment in the liquid; if the air is excluded, or only heated air has access to the solution, no lactous fermentation will take place, unless the proper ferment is added (Pasteur *). According to Béchamp, ordinary chalk contains in itself a ferment recognizable by the microscope; he found, when artificial carbonate of lime was used for the lactous fermentation, that the operation did not take place at all. The spontaneously-developed fermentation of saccharine juices containing nitrogen is sometimes lactous and sometimes vinous, but more frequently both together.

In order that a sugar solution may undergo the lactous fermentation there is added to it, at a temperature of from

---

* *Ann. Chim. Phys.*, 52.

20° to 40° C. (best about 30°), some putrid cheese or a suitable animal membrane, and a considerable quantity of chalk to neutralize the lactic acid as it is formed, because the presence of the latter hinders the progress of the fermentation by coagulating the caseine of the cheese. The whole is left for two or three weeks, when a crystalline deposit of calcium lactate is formed; if this is not removed it gradually redissolves, owing to the ensuing butyrous fermentation whereby butyric acid is formed, whose calcium salt is soluble; hydrogen is evolved at the same time. The reaction is illustrated by the equation:

$$\underset{\text{Lactic acid.}}{2C_3H_6O_3} = \underset{\text{Butyric acid.}}{C_4H_8O_2} + 2CO_2 + H_2.$$

Pasteur considers that the butyrous fermentation is excited by a peculiar infusoria. Slightly alkaline solutions are best suited to the development of the lactous fermentation, neutral liquids for the development of yeast (Pasteur). Lactous fermentation may be replaced by a conversion of the cane-sugar into acetic acid instead of lactic; this change is said to take place under the influence of the *Torula aceti* (Blondeau). According to Boutroux (*Compt. Rend.*, 1878, No. 9), the lactic ferment, and the fungus *Mycoderma aceti* which is associated with the acetification of alcohol, are identical, the function varying with the composition of the putrescent medium.

Vinous Fermentation.—The clear juice of saccharine plants, or any other solution of cane-sugar containing a suitable nitrogenous body, left to itself in contact with the air at a temperature of 20° to 24° C., becomes turbid after a few hours and gives off carbonic acid, the temperature of the solution rising at the same time; in from 48 hours to several weeks, according to the temperature and the nature of the nitrogenous matter present, the whole of the sugar is decomposed. The fermentation may take place at much lower temperatures than that given, down to 0°, but the action is then rendered very slow (Dubrunfaut). As soon as the evolution of carbonic acid is terminated, a substance previously suspended in the solution, is partly carried upwards by the adhering gas-bubbles, and partly falls to the bottom of the vessel; this insoluble substance is the *yeast.* The liquid, after the completion of the operation, contains, in the place of the sugar, alcohol, glycerin, and succinic acid mainly, together with traces of several other bodies. The formation of these products may be roughly represented by the following equations:

$$\underset{\text{Dextrose.}}{C_6H_{12}O_6} = \underset{\text{Alcohol.}}{2C_2H_6O} + 2CO_2, \quad (1)$$

and

$$49C_6H_{12}O_6 + 30H_2O = \underset{\text{Succinic acid.}}{12C_4H_6O_4} + \underset{\text{Glycerin.}}{72C_3H_8O_3} + 30CO_2. \quad (2)$$

—(Pasteur.*)

By far the greatest portion of the sugar is converted into alcohol and carbonic acid, only from four to five per cent. being transformed into other bodies. The yeast itself takes up from one to one and a half per cent. of the elements of the sugar in the form of cellulose and fat. According to Maumenè,† there is produced in the vinous fermentation small quantities of other alcohols, as butylic, amylic, and others, but no methylic. The gum, extractive, malic acid, and dextrin contained in fermenting liquids are not affected (Proust and Ventzke ‡). When a solution contains less than

* *Ann. Chim. Phys.*, 58.

† *Traité.*

‡ *Jour. Pk. Chemie,* xxv. 81.

four parts of water to one of sugar, the fermentation takes place imperfectly or not at all.

The exact *rationale* of the process of fermentation is a matter of some obscurity, though it is certain that the presence of albuminoid nitrogenous matter is essential, as well as a peculiar ferment; both of these are contained in yeast. The ferment is a fungoid growth, and is specifically different for the various classes of fermentation. The fungus *Saccharomyces cerevisiæ* appears to be the exciting agent for the vinous fermentation, while *Penicilium glaucum* performs the same office in the lactous fermentation; both are found in beer-yeast. Yeast produced in the alcoholic fermentation is capable of exciting the same change, under suitable conditions, in other saccharine solutions, air being necessary for the beginning of the operation, but not for its continuance.

Cane-sugar, previous to undergoing the vinous fermentation, is converted into a mixture of dextrose and levulose, in varying proportions, by the taking up of one molecule of water. This inversion is effected under the influence of a peculiar body contained in yeast or in the kernels of fruits, and called by Barth * *invertin*, who describes it as a white powder, soluble in water, and giving a precipitate with plumbic acetate; he gives the mean composition as

Carbon	43.90
Hydrogen	8.40
Nitrogen	6.00
Oxygen	41.47
Sulphur	.63

See also Hoppe-Seyler (*Ber. Chem. Gesell.*, iv. 810), Gunning (*Ibid.*, v. 821), and Donath (*Ibid.*, viii. 795).

* *Ber. Chem. Gesell.*, 1878, No. 5.

A solution thus wholly or partially inverted exhibits a levo-rotatory power with polarized light before and during the progress of the fermentation. It is undecided whether the inversion takes place, as with acids, according to the equation:

$$C_{12}H_{22}O_{11} + H_2O = C_6H_{12}O_6 + C_6H_{12}O_6,$$

or that the relative proportion of the glucoses formed varies from the above; the inversion is not due to the presence of succinic or any other acid.

According to Pasteur, the following represents, within narrow limits, the quantitative results of the vinous fermentation of cane-sugar:

100 parts of saccharose, equivalent to 105.36 parts inverted sugar, give—

51.11	parts	ethylic alcohol.
48.89	"	carbonic acid.
.67	"	succinic acid.
3.16	"	glycerin.
1.00	"	cellulose, fat, and extractive.

The glycerin and succinic acid are formed by the yeast, and not by any peculiar ferment.

Influence of Saline Matters on Fermentation.—Ammonium chloride precipitates yeast from a liquid, while potassium silicate and borax coagulate it. Maumené * gives the following, showing the action of salts on the vinous fermentation: one gramme of yeast was left for three days in contact with a solution containing thirty to forty grammes of salt. The results are approximate only—

1. *Fermentation more or less aided.*—Sulphates of potassium[1], sodium[14], magnesium[19], calcium[22], zinc[24], copper[26],

---

* *Traité*, vol. i.

aluminium[24]; chlorides of potassium[1], calcium[20], strontium[23]; phosphates of potassium[3], calcium[21], ammonium[18], sodium[15]; potassium formiate[6]; potassium tartrate[9] and bitartrate[10]; sodium lactate[17].

2. *Fermentation more or less retarded.*—Sulphates of iron[15] and manganese[18]; sodium sulphite[6]; nitrates of potassium[2] and ammonium[11]; chloride of barium[14]; iodide of potassium[4]; arseniate[6] and butyrate of potassium[7]; borate of sodium[8], Rochelle salt[9]. (The figures refer to the relative order of activity.)

3. *Inversion increased without fermentation being affected.*—Potassium nitrite, chromate, bichromate; sodium chloride, nitrite, acetate; ammonium chloride; mercuric cyanide.*

The Cellulosic Fermentation.—According to E. Durin,† cane-sugar is capable of breaking up, under the influence of a peculiar ferment and certain conditions, into levulose and cellulose, as shown by the following equation:

$$\underset{\text{Cane-sugar.}}{C_{12}H_{22}O_{11}} = \underset{\text{Cellulose.}}{C_6H_{10}O_5} + \underset{\text{Levulose.}}{C_6H_{12}O_6}.$$

During the progress of this fermentation the cane-sugar is transformed into levulose; in the simplest phase of the operation no gas is disengaged, but if the solution becomes acid carbonic acid appears, though acetic acid is principally formed. During the continuance of the fermentation a quantity of white clots are formed, which, following M. Durin, are pure cellulose. These, on being added to a sugar solution, can excite in it the same change by

* Compare results of Knapp, *Ann. der Chemie*, clxiii. 65; Dubrunfaut, *La Sucrerie Indigène*, xii.; Dumas, *Mon. Scientif.*, 1872; Kolbe and Meyer, *J. Pk. Ch.*, ix. 133.

† *La Sucrerie Indigène*, xi. 8.

which they were themselves formed; the temperature of 30° is the most favorable. Dextrose and mannite do not undergo this species of fermentation.

GENERAL CHEMICAL PROPERTIES OF SUGARS.

Action of Heat.—Some sugars containing water of crystallization—as dextrose, melitose, eucalyn, inosite, and trehalose—lose it at 100°, melezitose at 110°, and mycose at 130°. At somewhat higher temperatures the glucoses give up a further quantity of water and yield anhydrides analogous to mannitan. Thus,

$$\text{Dextrose } C_6H_{12}O_6 - H_2O = \underset{\text{Glucosan.}}{C_6H_{10}O_5}.$$

As the temperature is increased a number of indefinite bodies are formed, known as *caramel* and its derivatives. Submitted to dry distillation, the sugars are resolved into carbonic oxide, carbonic acid, methane, acetic acid, aldehyde, furfurol, acetone, liquid hydrocarbons, and a black coal.

Action of Oxidizing Agents.—The sugars are easily oxidized with powerful oxidizing agents, yielding products of simpler composition, as carbonic, formic, and oxalic acids. Glucose and levulose reduce salts of copper, silver, mercury, and bismuth quite readily. Haberman and Honig (*Chem. Centb.*, xiii. 119) claim that, by the action of *hydrated oxide of copper* on sugar solutions in the heat, (1) levulose, dextrose, invert, milk, and cane-sugars reduce to sub-oxide; (2) the reaction is very quick with levulose and invert-sugar, but less so with dextrose, while with cane-sugar it only begins after several hours' boiling, and even then the action is probably due to inversion; (3) the oxidation products are carbonic, formic, and glycollic acids, together with an amorphous body. By prolonged boiling with *nitric acid* saccharose yields mostly oxalic acid. At lower temperatures and with more dilute acids products are formed nearer in constitution to the sugars, as *mucic*, *saccharic*, *tartaric* acids, and sometimes racemic. The formation of the isomeric acids mucic and saccharic is illustrated by the equation:

$$C_6H_{12}O_6 + O_3 = H_2O + C_6H_{10}O_8$$

Tartaric acid is probably formed by the further oxidation of saccharic acid, racemic by the oxidation of mucic. Saccharose and dextrose, and most other sugars yield by this gradual oxidation only saccharic acid; lactose yields mucic acid principally, with a small quantity of saccharic; melitose gives saccharic mainly, with a small quantity of mucic acid.

The following table of Horneman* shows the relative quantities of tartaric and racemic acids formed by the oxidation of various carbohydrates:

100 parts will give	Parts of	
	Tartaric.	*Racemic.*
Lactose.........................	55.4	44.6
Gum.............................	63.0	37.0
Saccharose......................	59.7	40.3
Starch..........................	100.0	
Dextrose........................	100.0	
Levulose........................		100.0
Saccharic acid..................	72.6	27.4
Mucic "	?	100.0

Under the influence of *chlorine* and *bromine* some sugars yield two acids containing six atoms of carbon—*isodiglycoethylenic acid* $C_6H_{12}O_6$, and *gluconic* acid $C_6H_{12}O_7$.

* *Jour. Prak. Chemie,* lxxxix. 283.

The first is formed when a solution of bromine is made to act on milk-sugar; the second when a current of chlorine is passed through a dilute solution of cane-sugar or dextrose. Levulose and sorbite break up by the action of chlorine into glycollic acid.

Reactions with Acids.—Sugars form with acids compounds analogous to ethers, acting like polyatomic alcohols. Concentrated *nitric acid*, or a mixture of nitric and sulphuric acids, acts upon saccharine bodies, giving rise to nitro-substitution compounds in which the univalent radical NO_2 takes the place of an atom of hydrogen. Thus, in the case of saccharose, the product has the composition $C_{12}H_{18}(NO_2)^4O_{11}$. With inosite, *hex-nitro inosite* is produced, $C_6H_6(NO_2)^6O_6$. Isodulcite, dextrose, milk-sugar, and trehalose yield nitro compounds whose composition is not exactly known.

Sulphuric acid acts on cane-sugar much more strongly than upon the glucoses. A strong syrup of cane or milk-sugar mixed with concentrated sulphuric acid is immediately decomposed with strong intumescence, attended with an evolution of sulphurous acid gas and various volatile compounds, a black carbonaceous residue being left. Dextrose, under the same circumstances, gives without blackening, a sulpho-acid $C_{24}H_{48}SO_{27} = 4C_6H_{12}O_6.SO_3$, the reaction being precisely similar to that of organic acids with sugar. Phosphoric acid appears to act in the same manner.

Saccharides.—The *organic acids* yield, with sugars, ethereal compounds called saccharides. Berthelot* has produced this class of compounds by heating dextrose with various organic acids, such as acetic, butyric, stearic,

* *Ann. Chim. Phys.*, liv. 78.

but has found, as a general rule, that the number of molecules of water eliminated is one in excess of the number of molecules of the monobasic acid taking part in the reaction. So that the products obtained are ethers of glucosan, and not of glucose, as below:

$$(1)\quad C_6H_{12}O_6 + 2C_4H_8O_2 = C_6H_{10}(C_4H_7O)'O_6 + 2H_2O.$$
Dextrose. Butyric acid. Dibutyric glucose.

$$(2)\quad C_6H_{12}O_6 + 2C_4H_8O_2 = C_6H_8(C_4H_7O)''O_5 + 3H_2O.$$
Dibutyric glucosan.

By the action of tartaric acid on saccharose, dextrose, and lactose, according to the same chemist, entirely similar derived compounds are formed, which bear the relation to glucosan shown in the equation (2). Mannite also gives compounds likewise related to mannitan.

Action of Weak Acids—Inversion.—When cane-sugar is heated with dilute sulphuric acid or hydrochloric it is converted into dextrose and levulose:

$$C_{12}H_{22}O_{11} + H_2O = C_6H_{12}O_6 + C_6H_{12}O_6.$$

Melezitose yields two molecules of dextrose; melitose, one molecule of dextrose and one of eucalyn; and lactose, two molecules of galactose (Pasteur).

Action of Alkalies.—Dextrose is much more easily acted upon by caustic alkalies than saccharose. The decomposition of aqueous solutions of the glucoses takes place slowly in the cold, more quickly on heating, the liquid first turning yellow and then brown, yielding humus-like bodies. Dextrose thus treated gives glucic acid as the first product of the reaction. The sucroses $C_{12}H_{22}O_{11}$ are not attacked by dilute alkalies in the cold, and but slowly on heating; they are decomposed by boiling with concentrated alkaline solutions. When fused with caustic alkalies they yield oxalic acid.

Ammonia, in the form of gas or in aqueous solution, when allowed to act on the sugars and some other carbohydrates, is capable of forming compounds with them somewhat resembling gelatin, and containing in some cases from 14 to 19 per cent. of nitrogen. Dusart, by heating dextrose, lactose, and starch with aqueous ammonia in sealed tubes to 150° C., obtained nitrogenous substances which were precipitated by alcohol in tenacious threads, forming with tannic acid an insoluble, non-putrefying compound. As it has been observed that bone gelatin approximates in composition to an amide of the carbohydrates, the above facts are of considerable interest.

$$C_6H_{12}O_6 + 2NH_3 = \underset{\text{Gelatin.}}{C_6H_{10}N_2O_2} + 4H_2O.$$

It has also been observed that gelatin, when boiled with sulphuric acid, yields, among other products, sugars resembling the glucoses.

CHAPTER II.

CANE-SUGAR OR SACCHAROSE $C_{12}H_{22}O_{11}$.

Common Sugar—Crystallizable Sugar—Sucrose—Sucre de Canne, Fr.—Rohrzucker, Gr.

Occurrence.—Cane-sugar is widely diffused in the vegetable kingdom, being found more generally, and in greater quantities among the grasses. The sugar-cane, *Saccharum officinarum*, contains often more than twenty per cent. of sugar, unmixed, it is claimed, with any other sugar, when the plant is perfectly ripe. The following analyses of the cane are by O. Popp*:

	From Martinique and Guadaloupe.	From Cairo.	From Upper Egypt.
Water	72.22	72.15	72.13
Cane-sugar	17.80	16.00	18.10
Reducing do.	.28	2.30	.25
Cellulose	9.30	9.20	9.10
Salts	.40	.35	.42
	100.00	100.00	100.00

The stems of *Sorghum saccharatum* and *S. Holcus*, when quite ripe, contain 9 per cent. cane-sugar unmixed with fruit-sugar (Goessman); the unripe stems carry only starch and grape-sugar. P. Collier† has found in the

* *Zeit. für Chemie*, 1870, 328.

† *Report to the Commissioner of Agriculture*, 1879, and Aug. 1, 1880. Washington, U.S.A.

juice of different varieties of sorghum from 15.95 per cent. cane and .65 per cent. of grape-sugar, to 13.90 per cent. cane and 1.45 per cent. grape-sugar, when the canes are quite ripe. The juice from the stems of Indian corn or maize (*Zea mays*), according to the same authority, contained 12.00 per cent. cane-sugar and .68 per cent. grape-sugar. The nectar of flowers contains invert-sugar with a considerable proportion of cane-sugar, the latter amounting in the case of the fuchsia to three or four times the quantity of fruit-sugar (A. S. Wilson, *Chem. News*, xxxviii. 93).

Many fleshy roots carry considerable quantities of cane-sugar, notably those of *Angelica archangelica*, *Beta vulgaris*, *Chærophyllum bulbosum*, *Chicorium intybus*, *Daucus carota*, *Helianthus tuberosus*, *Leontodon taraxacum*, and others. The common beet (*Beta vulgaris*) averages from seven to eleven per cent. of cane-sugar, though in particular cases, owing to high cultivation, the amount has reached fourteen per cent.* The beet contains no other sugar besides saccharose. According to W. Stein,† eight per cent. of sugar is obtainable from the madder-root, though it contains fourteen per cent., partly uncrystallizable.

Cane-sugar occurs in the stems and trunks of trees, as the sugar-maple, *Acer saccharinum*, the sycamore, some species of *Betula*, in the vernal juice of *Juglans alba*, *Tilia Europæa*, and in several palms, especially *Saguerus Rumphii*, or the sago palm, and the *Cocos nucifera*, or cocoanut-tree. The leaves of many plants contain sugar. A. Petit found in vine-leaves .92 per cent. of

* Payen, *Compt. Rend.*, xl. 769 ; Schmidt, *Ann. der Chemie*, lxxxiii. 325.

† *Journ. für Prak. Chemie*, cvii. 444.

cane and 2.62 per cent. of grape sugar, and also the same bodies in cherry-leaves.

The sugar of fruits at the season of maturity is always cane-sugar, but by the influence of a peculiar ferment it may be partially or wholly converted into a mixture of dextrose and levulose, which is commonly called *fruit-sugar*. Ripe fruits thus sometimes contain only fruit-sugar, and at others a mixture of cane and fruit sugars. Buignet * gives in the following table the saccharine content of most of the common fruits, with the amount of acid present:

	Cane-Sugar.	Fruit-Sugar.	Acid.
Apricots	6.04	2.74	1.864
Pineapples	11.33	1.98	.547
English Cherries	.00	10.00	.661
Lemons	.41	1.06	4.706
Figs	.00	11.55	.057
Strawberries	6.33	4.98	.550
Raspberries	2.01	5.22	1.380
Gooseberries	.00	6.40	1.574
Oranges	4.22	4.36	.448
Peaches (green)	.92	1.07	3.900
Pears (Madeleine)	.36	8.42	.115
Apples	5.28	8.72	1.148
"	2.19	5.45	.633
Prunes	5.24	3.43	1.288
Grapes (hothouse)	.00	17.26	.345
" green	.00	1.60	2.485

The formation of cane-sugar in fruits is not prevented by the presence of acids (Buignet, *loc. cit.*) Cane-sugar is also found in melons and dates. Walnuts, hazel-nuts, bitter and sweet almonds contain only cane-sugar (Pelouze †), while the saccharine matter of others is a mixture of cane and fruit sugar. The sugar of common honey is levo-rotatory, and is composed of fruit-sugar, dextrose,

* *Ann. Chim. Phys.*, [3] lxi. 233. † *Compt. Rend.*, xl. 608.

and cane-sugar. The latter is found chiefly in the honey of the cells, and rapidly disappears on keeping, owing to an accompanying ferment. Cane-sugar is not found in healthy cereals and barley-malt ready formed, but is produced by the action of diastase and water in the crushed grain (Mitscherlich, Peligot, and Stein). The analysis of the manna from Sinai (from *Tamarix mannifera*) shows, according to Berthelot: *

55	per cent.	cane-sugar,
25	"	invert-sugar,
20	"	dextrin.

And that from Kurdistan:

61	per cent.	cane-sugar,
16.5	"	invert-sugar,
22.5	"	dextrin.

Preparation from Natural Sources.—For working on the small scale Marggraf recommends that the plant, reduced to as fine a state of division as is practicable, be treated with strong boiling alcohol, and the solution obtained filtered and allowed to cool, when the sugar crystallizes out. To obtain cane-sugar from fruits containing also uncrystallizable sugar, Peligot and Buignet † have adopted the following method: Add to the juice an equal volume of alcohol to prevent alteration, if it is to be kept any length of time before operating, and filter; saturate the filtrate with excess of milk of lime, and again filter. Boil the second filtrate, when a compound of cane-sugar and lime separates, which contains two-thirds of the total cane-sugar present. Filter, wash the precipitate well with water, diffuse it in water, and decompose with a stream of

* *Compt. Rend.*, liii. 583.

† *Ann. Chim. Phys.*, lxi. 233.

carbonic-acid gas. The solution filtered from the carbonate of lime is concentrated by heat (best in a vacuum) to a syrupy consistency, decolorized by bone-black, and mixed with strong alcohol until it becomes cloudy, when it is set aside to crystallize. If the solution, after treatment with carbonic acid, yields a turbid filtrate, solution of basic acetate of lead is added, the liquid refiltered, and the excess of lead removed from the second filtrate with sulph-hydric acid gas.

Physical Properties.—Cane-sugar when obtained by slow evaporation forms large, transparent crystals, but when the crystallization takes place rapidly they are much modified and striated. When a strong syrup is concentrated to the proper consistency, it sets, on cooling, to a solid mass of fine crystals, which, after being washed with a pure syrup, constitutes the *loaf-sugar* of commerce. Sugar crystallizes in the monoclinic system, the forms generally having hemihedral faces, but are often tabular.

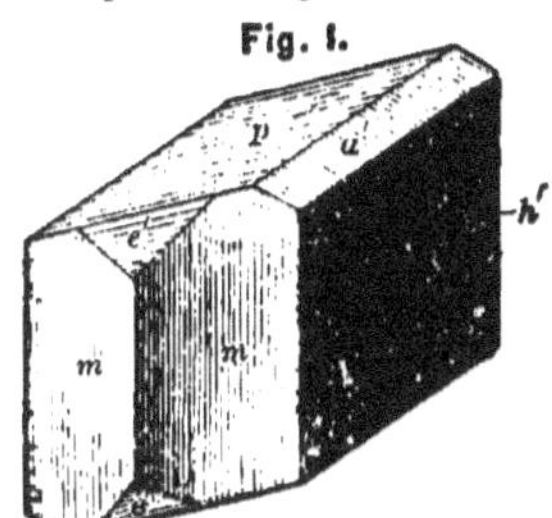

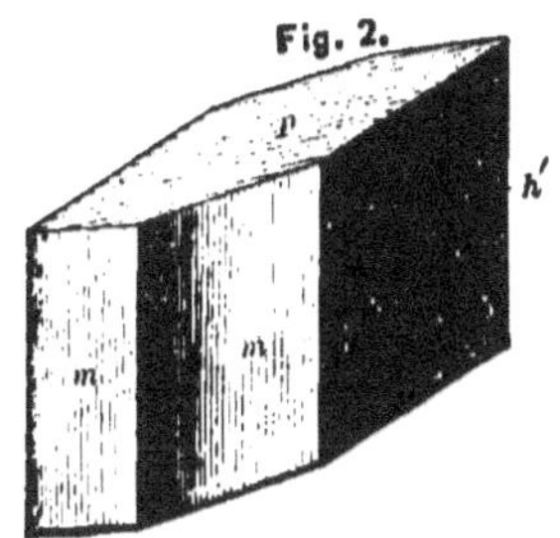

Figs. 1 and 2 show the crystallization of cane-sugar.

Axes : $a : b : c$	= .7952 : 1 : .70.
Angle of axes b and c	= 76° 44′.
Angles $p\ h'$	= 103° 30′.
$m\ m$ (on the side)	= 101° 32′.
$e'e'$ (above p)	= 99°.
$a'h'$	= 64° 30′.

(See Wolff, *Jour. Pk. Chemie*, xxviii. 129).

Ordinary forms, *m*, *p*, *h'*, *a'*, *e'*, $d^{1/2}$. Harder than any other sugar except lactose. Fig. 3 represents fine crystals of cane-sugar under a moderate magnifying power.

Fig. 3.

Cane-sugar exhibits phosphorescence when broken, or when a strong electric discharge is passed through it.

Specific Gravity.—1.593 (Joule and Playfair), 1.595 (Maumenè), 1.630 (Dubrunfaut), 1.580 (Kopp); the latter number, according to Gerlach,* who has carefully experimented in this direction, is the most correct—1.58046 at 17½°

* *Zeit. f. Rubenz.*, xiii. 283.

C. being the figure he obtained: melted barley-sugar 1.509 (Biot).

SPECIFIC ROTATORY POWER.—This constant as given by different authorities for the line D is:

	c.	[*a*] D.	
Arndtsen	77.394	67.02°	*Ann. Chim. Phys.* [3], 54, 403.
"	47.276	67.33°	" " "
Stefan	33.762	66.37°	*Wiener Akad.*, 52, ii. 486.
"	21.608	66.75°	" " "
Wild	30.276	66.42°	*Polaristrobom*, 1865.
Tuchschmid	27.441	66.48°	*J. Pk. Chem.* [2], ii. 235.
Calderon	19.971	67.08°	*Compt. Rend.*, 83, 393.
"	9.986	67.12°	" "
Girard and Luynes	16.350	67.31°	*Compt. Rend.*, 80, 1354.
Weiss	14.570	66.04°	*Wiener Akad.*, 69, iii. 162.
Oudemans	5.877	66.90°	*Pogg. Ann.*, 148, 350.

c = No. of grammes of material in 100 c. c. of solution. The discrepancies shown above are principally due to the different conditions as to concentration and temperature for the various series of experiments. Schmitz * gives a general formula when $c = 85.68$ to 10.40:

$$[a]\,D = 66.453 - .00123621c - .000117037c^2;$$

and for more dilute solutions:

$$[a]\,D = 66.639 - .0208195c + .00034603c^2.$$

According to Tollens, † when $c = 0$ to 18, and $c = 18$ to 69, the formulas are respectively:

$$[a]\,D = 66.8102 - .015553c - .00005246c^2.$$
$$[a]\,D = 66.386 + .015035c - .0003986c^2.$$

The deviation of the D ray for 1 mm. quartz is 21.67° (Broch).

For the transition tint (the mean yellow ray) the figure

* *Ber. Chem. Gesell.*, 1877, 1414.

† *Ibid.*, x. 403.

$[\alpha]j = 73.8°$ is the one generally given, and which correctly corresponds to the normal weights of the various saccharimeters using the transition tint (within narrow limits). The numerical relation of the rotations for the line D and the transition tint is variously given at—

$[\alpha]j = 1.13061\ [\alpha]D$ for quartz (Broch), and
$[\alpha]j = 1.129\ [\alpha]D$ (Montgolfier *),
1.0961,
1.034 (Weiss, *Wiener Akad.*, lxix. 157),

for sugar solutions.

Tollens (*Ber. Chem. Gesell.*, 13, 19, 2297) gives the rotatory power of cane-sugar in various solvents as follows:

10 per cent. solutions.
$[\alpha]D$ = water, 66.667°.
" + ethylic alcohol, 66.827°.
" + methylic " 66.628°.
" + acetone, 67.396°.

The temperature exercises no important influence on the rotatory power (see page 170).

Sugar is unalterable in the air. Specific heat, .301.

Action of Light.—Raoult (*Journ. Fab. Sucr.*, 1871) states that cane-sugar in solution enclosed in a sealed tube from which the air has been expelled by boiling, and kept for five months exposed to light, was found to have been half converted into glucose; a similar arrangement in the dark remained unaltered. Kreusler † asserts that if the air and germs are completely excluded in the above experiment no change takes place. This view is confirmed by Pellet ‡ and Motteu (*Ber. Belg. Akad.*, 1877).

* *Bull. Soc. Chim.*, xxii. 489.

† *Zeit. f. Anal. Chem.*, xiv. 197.

‡ *Jour. Fabr. Sucre*, 19, 5.

Composition.—Cane-sugar is composed of carbon, hydrogen, and oxygen:

	Equivalents.	Centesimally.
Carbon................	144	42.11
Hydrogen..............	22	6.43
Oxygen................	176	51.46
	342	100.00

Cane-sugar, whether obtained from the cane, beet, or any other source, is identical in every physical and chemical property, and in constitution.

Endosmose.—The endosmotic equivalent, according to Joly, is 7.25, but is not constant, depending on the quality of the membrane, though independent of the temperature (Schmidt, *Pogg. Ann.*, 102). It is proportional to the density of the solution.

Solubilitics.—Sugar is very soluble in water, the concentrated solutions having that peculiar consistency denominated *syrupy*.

H. Courtonne,* confirming the results of Berthelot and Scheibler, gives the solubility of saccharose at 12.5° C. and 45° C.:

12½°. 100 grms. of water dissolve 198.647 grms. sugar.
45°. 100 " " " 245.000 "

The saturated solution at

12½° containing 66.5 per cent.
45° " 71.0 "

The specific gravity of a sugar solution saturated at

* *Zeit. f. Rubenz.*, 1877, 1033.

17½° is 1.3272 to 1.330 (Anthon*); 1.345082 (Michel and Kraft†).

Brix‡ gives the following formula for calculating the amount of contraction produced by the solution of cane-sugar in water:

$$V = .0288747X - .000083613X^2 - .000002051X^3,$$

wherein X = the percentage of sugar dissolved; at the maximum, for a solution of 56.25 per cent. X is equal to .9946.

Cane-sugar is insoluble in ether and cold absolute alcohol; eighty parts of hot absolute alcohol take up one part of sugar, which it deposits on cooling.§ *Aqueous alcohol* dissolves it more readily. Scheibler‖ has calculated the following table from his experimental data on the solubility of cane-sugar in dilute alcohol of various strengths:

---

* *Zeit. f. Rubenzucker Ind.*, 1868, 615.

For full tables of solubilities at different temperatures and densities, see page 116 and the end of the volume.

† *Ann. Chem. Pharm.*, lii. 105.

‡ *Zeit. f. Rubenzucker Ind.*, 1854, 304; 1874, 1111.

§ *Ibid.*, xxii. 246.

‖ *Ber. Chem. Gesell.*, v. 343.

Per cent. absolute alcohol.	At 0° C.		At 14° C.		At 40° C.	
By volume	Sp. Gr at 17½° C.	Grammes in 100 c. c.	Sp. Gr. at 17½° C.	Grammes in 100 c. c.	Sp. Gr. at 17½° C.	Grammes in 100 c. c.
0	1.3248	85.8	1.3258	87.5	....	105.2
10	1.2991	80.7	1.3000	81.5	....	95.2
20	1.2360	74.2	1.2662	74.5	....	90.0
30	1.2293	65.5	1.2327	67.9	....	82.2
40	1.1823	56.7	1.1848	58.0	....	74.9
50	1.1294	45.9	1.1305	47.1	....	63.4
60	1.0500	32.9	1.0582	33.9	....	49.9
70	.9721	18.2	.9746	18.8	....	31.4
80	.8931	6.4	.8953	6.6	....	13.3
90	.8369	.7	.8376	.90	....	2.3
97.4	.8062	.08	.8082	.36	....	.5

On comparing this table with the one showing the solubility of cane-sugar in water (page 116), it will be seen that the water in mixtures of alcohol and water poor in alcohol, dissolves more sugar than it can *per se;* but for mixtures rich in alcohol the contrary is the case.

Sugar has a great tendency to form supersaturated solutions, especially when the temperature has been lowered. Contact with a solid body in a fine state of division at once determines a precipitation of the excess of sugar (Sostman, *Zeit. f. Rubenz.*, xxii. 837).

Action of Heat.—Pure cane-sugar heated to 100° C., even for a long time, is scarcely altered in the absence of watery vapor; in the presence of water a relatively considerable change takes place (Motteu). At 173° to 177° C. it melts without loss of weight to a clear liquid, which on cooling solidifies to an amorphous mass called *barley-sugar*, gradually becoming opaque and somewhat crystalline. If the fused mass is kept at this temperature for a long time it is altered, without loss of weight, into levulusan and dextrose, as:

$$C_{12}H_{22}O_{11} = C_6H_{10}O_5 + C_6H_{12}O_6.$$

Barley-sugar reduces less of copper oxide in the alkaline solution of tartrate of copper than does dextrose; sp. rotatory power,* 48°.

Cane-sugar heated above 180° degrees becomes brown, loses weight, and on cooling, if exposed to the air, absorbs more water than it lost, deliquesces, and behaves with alkalies like dextrose (Peligot). When heated for a long time from 210° to 220° it froths up, the brown color becomes darker, and a large quantity of water is given off containing traces of acetic acid and furfurol; when frothing has ceased the residue left is caramel mixed with some unaltered sugar and a bitter substance called *assamar*.† When the temperature is raised, more water is evolved, and an insoluble matter produced, which increases in quantity when the temperature is carried to 250° to 300°. This insoluble body is of complex composition, being composed of at least three distinct substances, *caramelene*, *caramelane*, and *caramelin*. According to Peligot,‡ on heating cane or grape sugar to 220°, and treating the residue obtained with alcohol, unaltered sugar and a bitter substance are dissolved out, and caramel remains behind, containing on the average, when dried at 180°, $C_{12}H_{18}O_9$, or two molecules less of water than cane-sugar. It is soluble in water, precipitated by baryta-water and subacetate of lead, not fermentable, and insoluble in alcohol.

When sugar is subjected to dry distillation, caramelization takes place with evolution of gases. The gas given off at first is nearly pure carbonic oxide, and afterward

---

* Tollens, *Ber. Ch. Gesell.*, x. 1403.

† *J. Pk. Chem.*, lxxxii. 148.

‡ *Ann. Chim. Phys.*, lxxvii. 154.

carbonic acid and marsh-gas make their appearance. An aqueous distillate forms, which holds a viscid oil and tar, besides acetic acid, acetone, and aldehyde. A voluminous, porous coal remains, constituting 32 to 34 per cent. of the sugar treated.

Inversion by Heat.—Water acts precisely as do dilute acids in converting cane into invert-sugar.

$$C_{12}H_{22}O_{11} + H_2O = \underset{\text{Dextrose.}}{C_6H_{12}O_6} + \underset{\text{Levulose.}}{C_6H_{12}O_6}$$

According to Gayon,* sugar solutions in sealed tubes invert in the cold, but much more rapidly when heated. See also Heintz (*Zeit. f. Rubenz.*, xxiv. 232), Berthelot (*Ann. Chim. Phys.*, lxxxiii. 106), and Pillitz (*Fres. Zeit.*, x. 456). A series of experiments made by Pellet† shows the effect of concentration and temperature in causing inversion. The table below gives the amount of invert-sugar formed during ninety-six hours' heating:

Sugar in 100 c. c.	At 25° C.	At 50° C.	At 75° C.
10 grm.	.5975 grm.	3.0216 grm.	8.8100 grm.
30 "	.5275 "	2.9200. "	7.1825 "
60 "	.1025 "	.6450 "	5.490 "
90 "	Trace.	.1500 "	3.9776 "

Maumenè,‡ as the result of experiments, which are supported by previously-made observations of Soubeiran, Buignet, and others, claims that invert-sugar is a substance of variable composition, the latter depending upon the facts as to whether the inversion takes place under the influence of heat or acids, the time, degree of heat, rela-

* *Compt. Rend.*, 1877, No. 10. † *Journ. Fabr. Sucre*, xix. 10.

‡ *Traité de la Fabrication du Sucre*, tome i. 118–137.

tive quantities of acid and sugar, and other circumstances; that it may, according to the above conditions, consist of dextrose, levulose, and an optically inactive sugar (isomeric with the two others) in all proportions; and that the mixture may show the most varying optical rotation. A solution of cane-sugar, according to Maumenè, on being submitted to progressively-increasing inversion, begins to lose its dextro-rotation, which is reduced to zero, after which a left rotation begins to appear, caused by the excess of levulose, which attains a maximum; the rotation then gradually decreases until zero is again reached, and then a plus or minus reading is shown; and finally there is a tendency to assume a permanent dextro-rotation. As bearing on the estimation of cane-sugar by Clerget's process, Maumenè allows that if strict attention is paid to the conditions laid down for the method (see page 136), the rotation of the inverted solution is constant, and hence no error from this source will be introduced.

Bechamp* attributed the inversion that cane-sugar undergoes in the presence of water to the influence of mould or fungi; but the later experiments of Clasen† seem to disprove this. The latter shows that water, acting as an acid, hydrates the cane-sugar, air being an important factor in the change. Nicol,‡ also, has proven that sugar is quickly and perfectly inverted when heated in sealed tubes to 130°–135°. A solution of sugar may be preserved for weeks in close vessels; but in a dilute syrup, exposed to the air and protected from dust, traces of altered sugar may be found in three days, which increase from day to day.

* *Compt. Rend.*, xl. 436. † *Journ. Prak. Chemie*, ciii. 449.

‡ *Amer. Chemist*, vi. 217.

Solutions of cane-sugar brought into intimate contact with the air alter very quickly. In an experiment where a solution of sugar of 10° B. was caused to flow over bits of broken glass in a cylinder open at both ends, at 19°C. it was found that traces of invert-sugar could be discovered after six hours. The alteration after this time went on with greater proportionate rapidity, so that scarcely any crystallizable sugar remained after thirty-six hours (Hochstetter*). When nitrogenous matter is present a few hours suffice for the above change. The best authorities admit that the formation of levulose and dextrose in inversion is simultaneous. Clasen (*loc. cit.*) states that a dilute solution of cane-sugar heated immediately after its preparation, nearly to the boiling-point of water for several hours, takes on no molecular change. According to Hochstetter, a solution of cane-sugar of 25° B., boiled in a dish for one, one and a half, and two hours, at 110° to 112°, underwent but slight inversion; but on passing air into the boiling solution the action took place with much greater rapidity. Lund ascribes this change to the carbonic acid present in the air.

ACTION OF ACIDS—INVERSION.

This change is produced in perfection by the action of dilute acids on cane-sugar solutions; the mineral acids act more quickly and powerfully than others. The change takes place at ordinary temperatures, but much quicker in the heat and as the acid is more concentrated. If the heating is long continued after the inversion is complete, coloration of the solution takes place, accompanied with the formation of various humus and ulmic compounds. If

* *Jour. für Prak. Chemie*, 1843.

a solution of sugar is heated with dilute sulphuric acid to a temperature below 100° C., ulmin and ulmic acid make their appearance, which, if the solution is brought to ebullition, are mixed with humin and humic acid; on the average not more than one-sixth of the sugar can be converted into these compounds, the rest remaining in solution as glucic acid, and, if the air has had free access, apoglucic acid is also present. The humus substances are produced at the boiling-point *in vacuo* (Mulder). Nitric and hydrochloric acids, as well as sulphuric acid, produce the humus decomposition of sugar; for every one part of these acids ten parts of oxalic, racemic, tartaric, citric, or saccharic acids, and sixteen parts of phosphoric, arsenious, arsenic, and phosphorous acids, are required to produce the same effect; the acid remains unaltered and may be recovered (Malaguti, *Ann. Chim. Phys.*, lix. 416).

The relation of various acids to the phenomena of inversion has been studied by A. Behr, * the results of whose experiments are—viz.:

I. *Effect of quantity of acid and concentration.*—For the same quantity of sugar the increase of inversion is by no means proportional to the increase in the amount of acid present; while it appears, on the other hand, for a fixed quantity of acid, that the inversion bears a direct relation to the concentration or amount of sugar present.

II. *Effect of temperature.*—Elevation of temperature has an enormous influence over the amount of inversion produced. Also, every acid has a specific temperature at which the alteration begins: this for sulphuric acid is 30° to 40° C.; for phosphoric acid, 40° to 50°; and for acetic acid, 70° to 80°.

---

* *Zeits. f. Zuckerind. des Deut. Reiches,* xxiv. 778.

III. *Effect of time.*—Probably the time during which the action takes place is in direct proportion to the amount of inversion.

IV. *Effect of the kind of acid.*—The following table shows the specific influence of different acids. The conditions of the experiments are the same in each case, the solutions being pretty concentrated, and the amount of acid is such that 100 parts of sugar have a quantity of acid chemically equivalent to one part SO_4H_2. In the columns headed 1, 2, 3, the time of the reactions and the temperatures were:

1. 13°–18° C., time 211 hours.
2. 19°–27° C., " 115 "
3. 25°–27° C., " 78 "

	Inversion, per cent.			Inversion relative to HCl; HCl = 100.		
	1.	2.	3.	1.	2.	3.
Acetic	.88	.97	1.29	1.2	1.3	1.6
Butyric	....	1.49	1.98	....	1.9	2.5
Isobutyric	....	1.56	2.05	....	2.2	2.5
Succinic	....	2.75	3.19	....	3.5	4.0
Malic	....	6.32	7.07	....	8.1	8.8
Citric	5.9	7.14	8.21	8.2	9.2	10.2
Formic	....	7.14	7.76	....	9.2	9.6
Lactic	7.38	8.10	7.99	10.2	10.4	9.9
Tartaric	8.19	10.41	11.10	11.4	13.4	13.8
Phosphoric	17.42	20.07	21.67	24.2	25.8	26.9
Oxalic	35.79	41.34	43.95	49.6	53.1	54.5
Sulphuric	60.52	64.68	67.91	83.9	83.1	84.2
Hydrochloric	72.10	77.84	80.68	100.0	100.0	100.0
Nitric	72.18	78.14	80.75	100.1	100.4	100.1

In the above experiments the inversion was estimated optically, the invert-sugar formed being calculated by the formula $p = \frac{200\,(100 - D)}{288 - t}$, in which

D = the polarization,
t = the temperature.

Lowenthal and Lenssen * have also experimented upon this subject, and some of their results differ from those given above; they are as follows:

1. The action is proportional to the quantity of cane-sugar present. 2. The action is slower the more dilute the solution up to a certain point, but beyond that the case is reversed. 3. The early stages of the reaction proceed more actively; all monobasic acids modify sugar in the same degree. Dubrunfaut † states that the amount of inversion by acids is directly as the square of the time, and for a complete inversion the time required is proportional to the quantity of acid; but, on the other hand, there is no simple relation to the quantity of sugar taking part in the reaction. According to Fleury,‡ using the same quantity of acid on varying quantities of sugar, the time required for complete inversion is constant; the results obtained by him are expressed by the curve

$$1 - y = K.f(a) - x,$$

in which K is a coefficient depending on the temperature and the nature of the acid, and (a) is a function of the amount of acid.

V. Lippman has investigated the inversion of cane-sugar by carbonic acid gas, and finds that a sugar solution saturated with the gas and polarizing 100°, after standing 150 hours showed a rotation of —44.2°, being completely inverted. The same authority also found the invertive action to be considerably increased by pressure, a solution at 100°,

---

* *Journ. Prak. Chemie,* lxxxv. 321.

† *Journ. Fabr. Sucre,* xiii. No. 21.

‡ *Compt. Rend.,* lxxxi. 823.

saturated with the gas and heated under pressure, being entirely inverted in from twenty to thirty minutes.

Bodenbender and Berendes* find that sulphurous acid inverts cane-sugar largely, especially in the heat, with the formation of sulphuric acid, and that the presence of citrates, lactates, formiates, and benzoates lessens this action; citric acid may entirely prevent it when the amount of the sulphurous acid is no greater than sufficient to saturate the base combined with the organic acid.

Hydrochloric acid in the cold forms only dextrose (Neubauer†). Acids of the aromatic series and salicylic acid invert strongly (Pellet and Pasquier‡).

Action of Sulphuric Acid.—Cold oil of vitriol forms a mixture with cane-sugar completely soluble in water without separation of carbon; if the solution is diluted, neutralized with chalk, and evaporated to dryness, it yields a dark-brown residue containing sulphur (Braconnot§). Mulder,‖ on heating cane-sugar with dilute sulphuric acid, found that *glucic acid* $C_{18}H_{18}O_9$ and *apoglucic acid* $C_{24}H_{26}O_{13}$ were formed. See also Richard (*Zeit. f. Rubenz.*, xx. 529).

By the action of sulphuric acid on sugar Grothe and Tollens¶ have produced a compound which they call *levulinic acid;* this acid has the formula $C_5H_8O_3$, and is produced by the decomposition of the levulose formed from the cane-sugar by inversion; the equation below represents the ultimate reaction:

$$C_{12}H_{22}O_{11} = C_6H_{12}O_6 + C_5H_8O_3 + CH_2O_2.$$

Dextrose. Levulinic acid. Formic acid.

* *Zeit. f. Rubenz.*, xxiii. 21.

† *Fres. Zeitschrift*, xv. 188.

‡ *Jour. Fabr. Sucre*, xviii. No. 33.

§ *Ann. Chim. Phys.*, xii. 189.

‖ *Ann. der Chemie*, xxxvi. 243.

¶ *Ibid*, clxxv. No. 1, 2.

In the heat sulphuric acid acts powerfully on cane-sugar, decomposing it, with evolution of sulphurous acid, and leaving a voluminous porous coal.

Hydrochloric acid gas is slowly absorbed by cane-sugar, which is converted into a brown product containing the elements of the acid; a concentrated solution of the gas acts violently and chars the sugar, with formation of ulmic acid.*

Phosphoric acid, when distilled with cane-sugar, produces formic acid and a volatile oil. *Oxalic acid* heated with an equal weight of cane-sugar, and the mixture distilled, yields formic and carbonic acids in small quantity, and a brown body similar to the *humin* of Mulder (Van Kerckhoff, *Journ. Pk. Chemie*, lxix. 48). *Tartaric acid* acting on cane-sugar at 100° forms *saccharoso-tetratartaric acid*, which reduces Fehling's solution and appears to contain modified saccharose (Berthelot). The acids *stearic*, *butyric*, *acetic*, and *benzoic*, heated with cane-sugar, furnish products resembling those obtained from dextrose under similar circumstances.

Action of Oxidizing Agents.—Nitric acid, or a mixture of nitric and sulphuric acids, acting on sugar, produces a nitro-substitution compound, *xyloidin*, which separates as a tough mass and inflames in contact with a red-hot coal. Dilute nitric acid gives rise to the formation of saccharic and oxalic acids. One part of cane-sugar with three parts of nitric acid of sp. gr. 1.25 to 1.30, heated to 50°, is entirely transformed into saccharic acid and water:

$$C_{12}H_{22}O_{11} + O_6 = 2C_6H_{10}O_8 + H_2O.$$

* Mulder, *Journ. Pk. Chem*, xxi. 203; xxxii. 331.

At a higher temperature oxalic acid is chiefly formed after long-continued heating with excess of acid.* By distillation with *peroxide of manganese* and sulphuric acid cane-sugar yields formic acid and a strongly-smelling oil. Boiling with *peroxide of lead* or *acid potassium chromate* gives formic and carbonic acids. Mixed with *chlorate of potassium*, and struck sharply or touched with a drop of oil of vitriol, cane-sugar explodes. PERMANGANATE OF POTASSIUM oxidizes sugar readily in acid solution, resolving it into carbonic acid and water. Maumenè has obtained two acids, *hexepic* $C_6H_{12}O_9$ and *trijienic* $C_3H_4O_5$, by the action of potassium permanganate on sugar. To prepare them, equal parts of sugar and the salt are dissolved in thirty to forty parts of water separately, and the solutions mixed in the cold, with agitation; heat is disengaged, and the manganese peroxide separates in a lump, and the clear, colorless solution contains the acids. Hexepic acid possesses a rotatory power equal to cane-sugar, and its solution is precipitated by acetate and subacetate of lead. Both acids form crystallizable salts. Hexepate of potassium is but little soluble; the trijienates of sodium, lead, and copper form small crystals. Maumenè has found these acids ready formed in a great number of plants, especially those yielding sugar.

Chlorine is absorbed by sugar, forming an odorous, deliquescent mass which gives off hydrochloric acid. When chlorine is passed into sugar solutions there is obtained, together with uncrystallizable compounds, a new acid free from chlorine, the barium salt of which is crystallizable. Malic acid is also said to be formed. Hlasiwetz and Haberman (*Ber. Chem. Gesell.*, iii. 486) have shown that chlorine

* See also Liebig, *Ann. der Chem.*, cxiii. 1.

acting on sugar gives rise to *gluconic acid* $C_6H_{12}O_7$. *Perchlorides* act on sugar in the same manner as chlorine, producing dark-colored products. Maumené * makes use of this fact as a basis for a qualitative test to detect cane-sugar and analogous substances. A drop of the liquid to be examined is placed on a strip of white merino previously steeped in a solution of stannic chloride, and dried. After the addition of the sugar solution the merino is warmed over a lamp, and the presence of saccharine or saccharoidal matter is indicated by the appearance of a black spot. *Iodine* and *bromine* act on cane-sugar in the same way as chlorine, with production of gluconic acids (Grieshammer, *Chem. Centb.*, No. 44). When equivalent quantities of potassium bicarbonate and iodine are added, one after another, to an aqueous solution of cane-sugar, iodoform is produced on boiling (Millon, *Compt. Rend.*, xix. 271).

OXYGEN, or air, and especially ozonized air, passed over dry cane-sugar or through the aqueous solution at common temperatures, gives rise to carbonic acid and water from a small part of the sugar, while the rest is unacted on. *Ozone* produces no change in a neutral aqueous solution of cane-sugar, but when carbonate of sodium is present the sugar is slowly but completely oxidized to carbonic and formic acids (Gorup-Besanez †). *Arsenic acid* oxidizes cane-sugar, a red color being developed owing to the formation of humus compounds. Elsner proposes the following as a qualitative test to detect cane-sugar: A solution containing one-thirtieth part of saccharose, heated in steam with a one per cent. solution of arsenic acid, in a

---

* *Traité.*

† *Ann. der Chem.*, cxxv. 211.

small dish, becomes red at the margin, and on evaporation yields a red spot.

With CUPRIC SALTS, sugar is slightly oxidized, the extent of the reaction varying with the conditions, such as whether the solutions are heated or not, the presence of caustic alkalies, etc. Cupric hydrate preserves its color when left to stand in the cold, or when boiled a short time with cane-sugar; but after longer boiling it gives up its water, turns brown, and is reduced to yellow cuprous oxide. If the solution contains a trace of alkali the hydrate dissolves immediately, and is then precipitated by the sugar. When cupric hydrate, washed with cold water, is boiled with cane-sugar solution and a little caustic alkali, the colorless liquid filtered from the precipitated cuprous oxide contains oxalic, carbonic, and acetic acids. Sugar boiled with aqueous cupric sulphate throws down metallic copper, while a quantity of cuprous salt remains dissolved (Vogel). A solution of equal parts sugar and cupric sulphate, and a sufficient excess of caustic alkali, retains its blue color unaltered in the cold for several days, and deposits a small quantity of red oxide only after some weeks. The reduction does not take place until after some time, even on boiling (Trommer *). When saccharose is boiled with cupric chloride, the liquid, on cooling, deposits cuprous chloride, if sufficiently concentrated. From cupric acetate a large quantity of cuprous oxide containing organic matter is thrown down, while a deliquescent sugar remains dissolved (Vogel †).

If a concentrated solution of cane-sugar is mixed with *cobaltic nitrate*, a small quantity of fused caustic soda added, and the solution boiled, a violet-blue precipitate is

* *Ann. Pharm.*, xxxix. 360. † *Schweigger's Journal*, xiii. 102.

formed. The presence of a very small amount of dextrose prevents the reaction.

Action of Alkalies.—Caustic alkalies, their carbonates, and the oxides of the alkaline earths all act more or less powerfully on cane-sugar. The oxides generally form compounds called *sucrates*, in which the sugar acts the part of an acid.

AMMONIA.—According to Laborde,* on passing a current of dry ammonia gas over perfectly anhydrous sugar, it becomes at first opalescent, and then takes on the waxy consistency described by Raspail; in the course of twelve hours it liquefies, and contains then 7.83 per cent. ammonia. Dextrose similarly treated liquefies very quickly and becomes colored, forming a crystalline compound. Cane-sugar heated with aqueous ammonia in sealed tubes for forty hours to 180° C. produces an insoluble black substance of undetermined composition.†

SODA AND POTASH.—Cane-sugar triturated with the fixed caustic alkalies, or strong solutions of them, is not colored brown, and this distinguishes it from dextrose. If cane-sugar be heated with caustic potash and a little water, the mass evolves hydrogen and is found to contain a large quantity of potassium oxalate (Gay Lussac). The mass, on distillation with sulphuric acid, yields carbonic, formic, and acetic acids and metacetone. Caustic alkalies and alkaline carbonates mixed with sugar diminish its rotatory power, not in proportion to the quantity of base present, but according to the concentration of the solutions. From such mixtures the sugar may be obtained with its original optical rotation by treatment with car-

* *Compt. Rend*, lxxviii. 82.

† Schutzenberger, *Ann. Chim. Phys.*, iv. 65.

bonic acid, the alkaline bicarbonates formed having no effect on the polarized ray (Sostman *). By boiling solution of cane-sugar for seventy-two hours with one-fiftieth part crystallized sodium carbonate, an acid black liquid is produced possessing levo-rotatory power (Soubeiran).

CAUSTIC LIME.—Bouchardat and Soubeiran have found that solutions of cane-sugar mixed with hydrate of lime exhibit greater stability, when boiled or long kept, than pure aqueous solutions. If a solution of sugar supersaturated with lime is allowed to stand for a year in a tight bottle, the excess of lime contains neither oxalic nor malic acids. After removing the dissolved lime, evaporating to dryness, and redissolving in alcohol, cane-sugar crystallizes out from the alcoholic solution, while *melassic* and *saccharic* acids, and uncrystallizable sugar remain in the mother liquor (Brendecke). An intimate mixture of one part cane-sugar and three parts quicklime, heated, produces a violent reaction, acetone and metacetone being evolved. On distillation of sugar with caustic lime there are produced acetone, metacetone, isophorone, and marsh-gas, with small quantities of carbide of ethylene.

SUCRATES.

Potassium Sucrate.†—$C_{12}H_{21}KO_{11}$? is formed as a gelatinous precipitate by adding caustic potash to a strong solution of sugar in alcohol. It is white, friable (Sostman says it cannot be dried), and translucent; melts at 100° to a viscid liquid having an alkaline, not sweet taste;

* *Zeit. f. Rubenz.*, xxii. 173.

† Authorities: Peligot (*Ann. Chim. Phys.*, [2] lxvii. 113; *ibid.*, lxxiii. 103; *ibid.* [3] liv. 377). Soubeiran (*Journ. de Pharm.* i. 469). Berthelot (*Ann. Chim. Phys.*, [3] xlvi. 173).

completely decomposed by carbonic acid, the sugar being recovered unaltered.

Potassium Hydric Sucrate.—To a hot saturated solution of sugar an equal bulk of strong nitric acid is added, and the mixture kept warm until the evolution of gas has ceased, when it is boiled. The liquid is then divided into two equal parts, one of which is neutralized with caustic potash and added to the other, when an abundant precipitation of the sucrate takes place; this, if colored, may be purified by filtration over animal charcoal, evaporation, and recrystallizing (Bayley, *Chem. News*, xliii. 110).

Sodium Sucrate ($C_{12}H_{21}NaO_{11}$?)—Similar in all respects to the potassium compound.

Calcium Sucrates.*—Lime combines with saccharose in different proportions, forming combinations whose chemical constitution is mostly well marked. The quantity of lime dissolved by sugar solutions depends on their density and the temperature at which the solution takes place; this for 100 parts of sugar varies, under these circumstances, from 23 to 55 parts. When excess of lime is agitated with a sugar solution, saturation takes place but slowly, and only when the quantity of the base is at least twice as great as the solution will take up. Strong solutions (above 30 per cent.) become gummy and solidify, while with more dilute solutions monobasic sucrate is formed; but this is capable of taking up an additional quantity of lime, greater in proportion as the solution is more concentrated. Cane-sugar solutions of 40 per cent. dissolve 26.57

* Authorities: Soubeiran (*Journ. de Pharm.*, i. 469). Peligot (*Compt. Rend.*, xxxii. 333; *Ann. Chim. Phys.*, [3] liv. 377; *Compt. Rend.*, lix. 930). Berthelot (*Ann. Chim. Phys.*, [3] xlvi. 173). Pelouze (*Compt. Rend.*, lix. 1073). Boivin et Loiseau (*Compt. Rend.*, lix. 1073; *ibid.*, lx. 164, 454; *Ann. Chim. Phys.*, [4] vi. 203). Horsin-Deon (*Bull. Soc. Chim.*, 1871, xvi. 26; *ibid.*, xvii. 155).

parts of lime to 100 parts of sugar; solutions of 20 per cent. take up 23.15 parts; and 5 per cent. solutions dissolve 18.06 parts (Peligot).

Solution of calcic sucrate has a bitter and alkaline taste. The specific rotatory power of the sugar combined with lime is less than in the free state (see page 176). On neutralization with acid the rotatory power is restored, even if the solution of the sucrate has been heated to 117.5°, but not if heated higher (Dubrunfaut). A solution of calcic sucrate considerably diluted forms a gelatinous mass on heating; on cooling, or the addition of sugar, the solution is cleared up.

According to Bodenbender,* the aqueous solution of sucrate of lime dissolves certain metallic oxides in the presence of excess of the sucrate. The following table represents the amount of the oxides taken up by sucrate solution of various strengths. A is a solution containing in one litre 418.6 grm. sugar and 34.3 grm. lime; B contains 296.5 grm. sugar to 24.2 grm. lime; and C 174.4 grm. sugar to 14.1 grm lime:

	A.	B.	C.
MgO..........Grammes.	.30	.24	.22
Al_2O_3 .... "	1.35	.32	.19
Fe_2O_3 .... "	6.26	4.71	3.08
Mn_2O_3.......... "	.50	.37	.32
Cr_2O_3 .... "	1.07	.56	.20
CoO............ "	1.56	1.00	.59
NiO .... "	.29	....	....
ZnO............ "	....	....	.24
CdO............ "	.22	....	.48
CuO............ "	10.26	5.68	3.47

The solutions, on standing, deposit lime and the oxide.

* *Zeits. f. Zuckerind. Deut. Reiches*, 1865, 851–860.

An aqueous solution of lime sucrate dissolves recently-precipitated phosphate and carbonate of lime.

Only dilute solutions of the sucrate become turbid on exposure to the air; carbonic acid gas slowly but completely precipitates the base, yielding the sugar unaltered. When the solutions have been prepared in the cold no traces of invert-sugar can be detected by boiling with alkaline solution of oxide of copper. According to Hochstetter, even if the solution is boiled on the open fire for two hours till the mass begins to thicken and char, the unburnt portion still yields the sugar unaltered.

Monobasic Sucrate. $C_{12}H_{22}O_{11}$ CaO.—Prepared by adding 85 per cent. alcohol to a concentrated solution of sugar containing excess of lime. It is a white precipitate, drying to a brittle resin, which deflagrates after drying and dissolves easily in cold water; the solution, when heated, deposits tribasic sucrate and sets free some sugar.

$$3(C_{12}H_{22}O_{11}\ CaO) = C_{12}H_{22}O_{11}\ 3CaO + 2C_{12}H_{22}O_{11}$$

(Peligot).

S. Benedikt* prepares this compound by adding magnesium chloride to a sucrate containing excess of lime, filtering, and treating the filtrate with excess of alcohol; the precipitate produced is washed with warm 60 per cent. alcohol. Dried at 100°, the deposit has the formula $C_{12}H_{20}CaO_{11}$; dried in a vacuum at ordinary temperatures, two molecules of water are retained.

Bibasic Sucrate. $C_{12}H_{22}O_{11}$ 2CaO.—Boivin and Loiseau (*loc. cit.*) have obtained this compound (1) by agitating finely-divided hydrate of lime with a solution of cane-sugar, and cooling to 0°; (2) by treating the tribasic sucrate with

---

* *Bull. Soc. Chim.*, xx. 279.

sugar and lime; (3) by precipitating in the cold, by alcohol of 65 per cent., a solution of lime and sugar, and boiling. Water decomposes this sucrate into the tribasic salt and cane-sugar. The compound $C_{12}H_{22}O_{11}$ 2CaO $\frac{1}{2}H_2O$ is said to be obtained by precipitating a solution of sugar-lime by alcohol.

Sesquibasic Sucrate. $2C_{12}H_{22}O_{11}$ 3CaO.—This is always formed when a solution of sugar with excess of lime is boiled, or set aside at ordinary temperatures; the compound may be obtained as a white amorphous gum by evaporating the filtrate in an atmosphere of carbonic acid gas. It is a transparent, resinous or granular, white, friable mass, which deflagrates and readily dissolves in cold water; insoluble in strong and weak alcohol, but soluble in an alcoholic solution of cane-sugar.

Tribasic Sucrate. $C_{12}H_{22}O_{11}$ 3CaO. $C_{12}H_{22}O_{12}$ 3CaO.3H_2O.*—This separates as a mass resembling coagulated albumen, when a sugar solution containing excess of lime is heated and filtered. It is soluble in 100 parts of cold water, the solution when heated depositing half the quantity dissolved; it is readily soluble in sugar-water (Peligot, *loc. cit.*)

Sexbasic Sucrate. $C_{12}H_{22}O_{11}$ 6CaO.—According to Horsin-Deon, a salt of the above composition is obtained by treating the tribasic sucrate with alcohol.

Sucro-carbonates of Lime.†—When carbonic acid gas is passed into sugar-water mixed with lime, the gas is absorbed, and if the liquid is sufficiently dense a gelatinous precipitate is formed after a time; by the continued action

---

* Lippman, *S. Neue Zeit.*, iv. 148.

† Dubrunfaut, *Compt. Rend.*, xxxii. 498; Boivin-Loiseau, *Bull. Soc. Chim.*, xi. 345; Horsin-Deon, *ibid.*, xv. 22; xix. 65.

of the gas the precipitate is decomposed and all the lime thrown down. If the solution is heated the compound is likewise decomposed, but some lime remains in solution. The body thus formed, having the formula $3CO_2Ca.C_{12}H_{22}O_{11}$ $3CaO\ 2H_2O$, is the *hydrosucro-carbonate of lime* of Boivin and Loiseau, and may also be obtained by the action of carbonic acid on the sexbasic sucrate. According to Horsin-Deon, the compound $3CO_2Ca.C_{12}H_{22}O_{11}\ CaO\ 2H_2O$ is produced under other circumstances when the proportions of water, lime, and sugar are different from the above. The composition of the sucro-carbonate, however, varies with the temperature, density of the solutions, and with the varying proportions of sugar and lime; the quantity of carbonic acid absorbed may range from 4.4 per cent. to 16.28 per cent. Bondonneau * considers the sucro-carbonate to be only calcic carbonate in a gelatinous condition, and soluble in sucrate of lime.

Sucrate of Baryta. $C_{12}H_{22}O_{11}\ BaO$.—Prepared by adding to sugar solution, baric hydrate or sulphide:

$$C_{12}H_{22}O_{11} + 2BaS + H_2O = BaO\ C_{12}H_{22}O_{11} + BaSH_2S.$$

It consists of small nacreous crystals resembling boracic acid, of a caustic taste and alkaline reaction; after drying *in vacuo* it does not give off water at 200° F. Decomposed by carbonic acid, it gives up the sugar unaltered. Soluble in 47.6 parts of water at 15°, and 43.5 parts at 100°. Insoluble in wood-spirit and alcohol. The formula $C_{12}H_{22}O_{11}$ $2BaO$ has been assigned to this compound by Peligot, Stein, and others.

Sucrates of Lead.† (*a*) *Bibasic* $C_{12}H_{18}Pb_2O_{11}$.—This compound is formed when finely-divided litharge is boiled with

---

* *Bull. Soc. Chim.*, xxiii. 3.

† Boivin et Loiseau, *Compt. Rend.*, 1865, 60.

a solution of sugar, or when ammonia is added to a solution of sugar mixed with neutral acetate of lead; it is insoluble in cold water and alcohol, and soluble in boiling water, crystallizing out, on cooling, in nodules or needles. A solution of the tribasic sucrate left to stand deposits the bibasic salt, sugar being set free. (*b*) *Tribasic* $C_{12}H_{16}Pb_3O_{11}$.—Prepared by adding caustic soda or potash to a solution of acetate of lead and sugar, taking care not to have an excess of either of the bodies taking part in the reaction; or by mixing a solution of calcic sucrate with a boiling solution of acetate of lead. It is a white powder, insoluble in cold and but little soluble in boiling alcohol, but easily soluble in solutions of acetate of lead, caustic alkali, or cane-sugar. Metallic lead is attacked by cane-sugar solutions.

Sucrate of Strontia is formed by adding the hydrate to sugar-water.

Ferrous Sucrate $C_{12}H_{22}O_{11}$ FeO.—When metallic iron is partially immersed in a sugar solution it rapidly corrodes. The red-brown solution produced yields, on evaporation, a tasteless, insoluble residue corresponding to the above formula in composition. It is insoluble in alcohol, acted on by ammonium sulphide, and not by alkalies and their carbonates. Sugar solutions do not dissolve ferrous oxide, and have but slight action on ferric oxide. Ferric hydrate is dissolved by a solution of sucrate of lime, a reduction to protoxide taking place. By evaporation a double salt of the following composition is obtained:

$$FeO2CaO\ C_{12}H_{22}O_{11}\ 3H_2O.$$

Sucrates of Copper.*—Copper in partial contact with the air dissolves in sugar-water. Cupric carbonate is

* Barreswill, *J. de Pharm.*, iii 7, 29.

readily soluble in the same. A concentrated solution of cane-sugar and cupric sulphate, on standing, deposits a bluish-white precipitate containing $SO_4Cu\ C_{12}H_{22}O_{11}\ 4H_2O$.

The Double Sucrate of Lime and Copper, $CuO\ CaO\ C_{12}H_{22}O_{11}\ 3H_2O$, is obtained by evaporating a solution of calcic sucrate in which cupric oxide has been dissolved. It is crystallizable and soluble in cold water, forming a blue liquid.

Sucrate of Magnesia is formed by dissolving the hydrate in sugar-water. All of the magnesia is deposited from the solutions on standing.

Hydrate of alumina is slightly soluble in sugar solution. Oxide of zinc and silica are insoluble. Common metallic zinc in contact with iron is dissolved readily; but very small quantities of pure tin, zinc, mercury, or silver are dissolved under the same circumstances (Gladstone*).

COMBINATIONS OF CANE-SUGAR WITH NEUTRAL SALTS.†

With Sodium Chloride, $(C_{12}H_{22}O_{11})^2\ NaCl^6(H^2O)$ (Maumené), $C_{12}H_{22}O_{11}\ NaCl$.—This compound is deliquescent and affected by heat much in the same way as cane-sugar. It has a sweet saline taste, and the sugar retains its rotatory power unaltered. Ch. Violette gives the formula $C_{12}H_{20}$-$NaClO_{11}$, and considers it a product of substitution. If ether is added to an alcoholic solution of this body, an oleaginous layer separates, which deposits, little by little, crystals corresponding to the formula:

$C_{12}H_{22}O_{11}\ NaCl\ 2H_2O$ (Gill, *loc. cit.*)

Gill, when experimenting with cane-sugar mixed with 1, 2, 3, or 4 molecules of chloride of sodium, found the crystals

* *Journ. Chem. Soc.*, vii. 195.

† Gill, *Journ. Chem. Soc.*, [2] ix. 269.

were always of variable composition. He obtained a few crystals having the composition:

$$2C_{12}H_{22}O_{11}\ 3NaCl\ 4H_2O.$$

With **Ammonium Chloride** a crystalline compound of cane-sugar may be formed containing NH_4Cl, but the composition is not invariable.

With **Potassium Chloride,** $C_{12}H_{22}O_{11}$ KCl^3, crystallizes isomorphous with cane-sugar, and is not deliquescent. Gill and Maumenè were not able to obtain this combination of invariable composition.

With **Bromide of Sodium,** $C_{12}H_{22}O_{11}$ BrNa $1\frac{1}{2}H_2O$, crystallizes with difficulty and contains varying amounts of water.

With **Iodide of Sodium,** $C_{12}H_{22}O_{11}$ $3NaI.3H_2O$, crystallizes well in the monoclinic system, and the rotatory power of the cane-sugar contained is not altered. The composition is constant, no matter in what proportions the components are mixed (Gill).

Lithium chloride, bromide, and *iodide* do not form definite compounds with cane-sugar. Acetate, nitrate, and phosphate of sodium also do not appear to combine with cane-sugar.

With Borax, $3C_{12}H_{22}O_{11}$ $Na_2B_4O_7.5H_2O$.—When borax is dissolved in solution of sugar and the liquid evaporated, the salt first crystallizes out. On precipitating the mother-liquor with alcohol a glutinous liquid is thrown down, which, after solution in a small quantity of water, and precipitation with alcohol, yields a compound of the above composition (Sturenberg, *Archiv. der Pharm.*, xviii. 27).

MELASSIGENIC ACTION OF SALTS AND ORGANIC MATTERS ON SUGAR IN SOLUTION.

Melassigenic action consists in the formation of molasses, which is a residue of cane-sugar solutions from which all sugar capable of crystallizing has been obtained. The loss of crystallizing power through the action of salts may occur in two ways: (a) *either by an invertive action causing the cane-sugar to be transformed into invert-sugar*, or (b) *by a specific effect, different for each salt, whereby they retain the sugar in solution without altering its chemical constitution.* The lowest molasses of commerce, from which all sugar has been crystallized that it is practicable to get, will retain from 25 to 30 per cent. cane-sugar to about an equal quantity of invert-sugar.

A. **The Invertive Action.**—Some neutral salts have the power of inverting cane-sugar, either by their decomposition, setting free acids or forming acid salts, or by a specific agency. Béchamp * has made some experiments on this subject, but his results should be received with some caution, from the fact that the sugar solutions upon which he worked were in contact with the air from seven to eight months, by which mould may have been formed, and the inversion ascribed to the salts have been caused by the presence of fungi. Further, Béchamp measured the inversion by the lowering of the optical rotation; but as some of the salts themselves have a similar action on the polarized ray, the optical means cannot be relied upon for the purpose to which it was applied. The following are Béchamp's results: Aqueous solutions of sugar mixed with zinc sulphate, plumbic nitrate, monophosphate, or

* *Ann. Chim. Phys.*, liv. 28.

arseniate of potassium, or with a large quantity of mercuric chloride, lose their rotatory power partially or entirely by standing at ordinary temperature, and occasionally acquire a rotation to the left, without formation of mould. A sugar solution containing one-fourth of its weight of fused chloride of zinc or calcium hardly decreases in rotatary power in standing nine months, or when heated for an hour to 50°. The presence of small quantities of corrosive sublimate, zinc nitrate, and neutral or acid potassium sulphate prevents the formation of mould. Most other salts, as well as nitric and arsenic acids, do not hinder the formation of mould in sugar solutions, and in general the decomposition from this cause goes on more rapidly in their presence. If cane-sugar solutions are mixed with neutral or acid sodium sulphate and one drop of creosote, no appearance of decomposition takes place on standing; but the growth of fungi having once commenced, creosote has no power to arrest it. So far Béchamp. W. L. Clasen * has more recently made similar researches to those of the French chemist, in which he has avoided the sources of error inherent to the method of the latter. Fehling's copper test was used in connection with the saccharimeter to estimate the presence of invert-sugar, and the solutions were never allowed to stand more than five days, precluding the possibility of mould formation. The following are the conclusions based on his experimental results: (1) Some salts at ordinary temperatures hinder the formation of invert-sugar, as sulphate of lime, ammonic chloride, and potassium nitrate; others, as magnesium sulphate, weaken the agency of water in inverting, though they are not entirely able to prevent it. (2) If

* *Journ. f. Prak. Chemie*, ciii. 449; *American Chemist*, iv. 89.

cane-sugar solutions mixed with certain salts, after standing several days at ordinary temperatures, be heated to 88°, ordinarily a proportionally strong inversion takes place. This is the case with sulphate of lime, nitrate of potash, and sulphate of magnesia. Water containing sulphate of lime and ammonic chloride shows the strongest reaction in consequence of the formation of an acid salt. (3) Sugar solutions mixed with salts and heated to 88° immediately after preparation, indicated inversion only in the case of sulphate of lime and ammonic chloride. (4) The assumption of Béchamp that some salts, through their "personal" influence, can convert cane-sugar into invert-sugar without formation of mould, seems to be just.

According to Berthelot,* dry cane-sugar is not altered by being heated to 100° for several hours with NaCl, $SrCl_2$, or $BaCl_2$; but the addition of a small quantity of water causes inversion more abundantly than it would have in the presence of water alone; the same transformation takes place more quickly with ammonic chloride and a little water, the mass being blackened. Sodium chloride and fluor-spar do not seem to have the same effect.

The researches of Pellet,† upon the invertive action of glucose and salts, made under different conditions as regards time and temperature, give the following results: As regards *time*, glucose forms quicker the more dilute the solutions; *heat* increases the quantity of invert-sugar, and the action is stronger in dilute than in concentrated solutions; *glucose* aids the inversion the more as the quantity of it is greater—the action is *nil* in saturated solutions; *salts*—the inorganic salts have a much greater action at 50° to 60° than at ordinary temperatures; it is also greater

* *Ann. Chim. Phys.*, xxxviii. 57. † *Journ. des Fabricants*, xviii. No. 10.

with dilute solutions; nitrate of calcium acts more energetically than the chloride, and ammonic nitrate gives a powerful inversion. 100 c.c. of water, 10 grammes of sugar, and 5 grammes of the ammonia salt, heated half an hour, give a complete transformation of the sugar into invert-sugar.*

Durin found that the presence of invert-sugar in a solution of cane-sugar caused no inversion at 70° to 75° C. when the alkalinity is maintained at .001 of CaO; on heating, however, to 75° to 114° C., the solution becomes faintly acid, inversion begins and goes on until complete change of the sugar is effected; if the solution is kept alkaline no change takes place. The presence of invert-sugar is not necessary, the change taking place on formation of acids.†

B. **Action on Crystallization.**—The effect that mineral and organic salts have on the crystallizing power of cane-sugar in aqueous solutions has engaged the attention of many chemists, and the results obtained by them differ in important particulars both as to whether the specific melassigenic action exists at all, and as to the special effect of the different salts in this direction. The work done upon the subject has had reference principally not to pure sugar solutions, but to the juices and molasses derived from the beet; therefore due regard should be paid to this fact in the case of similar solutions from the cane, as the conditions are quite different. Beet and cane molasses contain,

---

* Bodenbender denies the completeness of the change, and ascribes it to the presence of free acid.

† The above results are quite in conformity with the experience of the author, who finds that a perfectly neutral or slightly alkaline solution of raw sugar, on being heated for some hours, invariably develops acidity with consequent inversion. The acid is probably formed by the decomposition of invert-sugar or impurities present.

on the average, when all the sugar that will crystallize has been obtained:

	Beet-molasses.	Cane-molasses.
Cane-sugar	55.00	35.00
Organic matters not sugar	13.00	10.00
Water	20.00	20.00
Glucose	Trace.	30.00
Ash	12.00	5.00
	100.00	100.00

The mineral salts in the two are very different in character, consisting, in the case of cane-molasses, in large part of lime salts of organic and mineral acids, with a comparatively small portion of alkaline salts, which are mostly chlorides and sulphates; in beet-molasses the salts are largely those of potassium combined as chloride, sulphate, and nitrate, or with organic acids. The organic matters associated in the two types are also as varied in kind, and the further influence of the large amount of glucose in cane-molasses renders still greater the essential dissimilarity in the two products. Still, to a certain extent and with the proper allowances, what is true of the action of salts on crystallization in the beet-sugar manufacture is also true with the products of the cane.

It has been widely asserted that the melassigenic effect is purely molecular and physical. Champion and Pellet,* as the result of an elaborate series of experiments on the subject, conclude that the action depends—

1. *On the influence of the active body on the solubility of the cane-sugar.*

2. *On its influence upon the boiling-point of the solution.*

---

* *Sucrerie Indigène,* xii. 210, 223, 257.

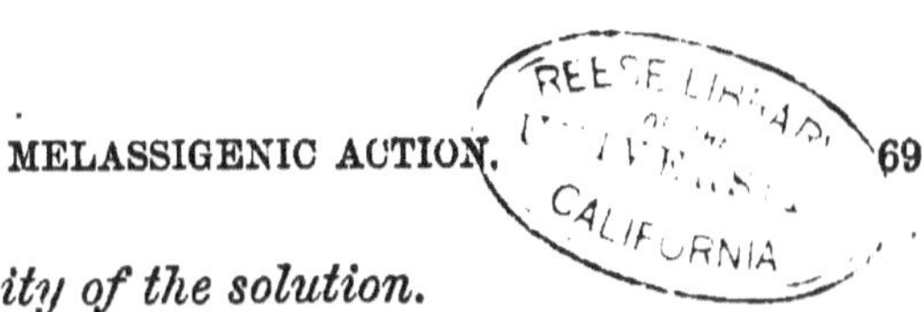

3. *On the viscosity of the solution.*

These conclusions are supported by other chemists.

Recently Gunning* has proposed a different explanation, and one which probably accounts for the phenomena to a great extent. His views, founded on experimental data, are that the saccharose contained in molasses (beet ?) exists for the most (nine-tenths of total) part in the form of a chemical combination of double salts, in which the cane-sugar is combined with organic compounds containing a mineral base, these double compounds being non-crystalline; if the theory is correct it offers an explanation of melassigenic action from a purely chemical point of view. Gunning finds that nearly all organic potash salts are capable of combining with sugar, though this property is not shared by the sodium salts for the most part. The following salts may, however, be excepted: sodium formiate and acetate, potassium phosphate and nitrate, sodium carbonate and barium chloride. The fact that the sugar in cane-molasses, as shown by the foregoing analysis (page 68), may be reduced to a lower quantity relative to the impurities, by crystallization, seems to support Gunning's views, as there is a comparatively small quantity of potassium salts in cane-juice and its products.

A. Marschall † has made a valuable series of experiments upon the influence of salts on the crystallizing power of sugar. He enclosed sugar in a sealed tube with a quantity of various salts, the amount of water present being less than half that of the sugar, and, therefore, not enough to dissolve all of it at ordinary temperatures; the tubes, after filling, were warmed until the sugar was in solution, and then allowed to rest in a cool place from 17 to 21 days,

* *Stammer's Jahresb.*, xvii. 181. † *Journ. Chem. Soc.*, [2] ix. 457.

when the sugar crystallized out. The mother-liquors were then examined, the sugar and salt contained being estimated. If the given salt prevented crystallization, the solution would contain a quantity of sugar less than the normal (2 of cane-sugar to 1 of water) for a saturation solution in the cold. The results were thus classified:

(*a*) NEGATIVE MOLASSES-MAKERS, or bodies which diminish the solvent power of water for cane-sugar, are: sodium sulphate, nitrate, acetate, butyrate, valerate, and malate; magnesium sulphate, nitrate, and chloride; and calcium chloride and nitrate.

(*b*) INDIFFERENT BODIES—without influence on crystallization, are: potassium sulphate, nitrate, chloride, valerate, oxalate, and malate; sodium chloride, carbonate, oxalate, and citrate, and caustic lime.

(*c*) POSITIVE MOLASSES-MAKERS are: potassium carbonate (saline coefficient .38), acetate (.9), butyrate (.9), and citrate (.6). The action of the negative molasses-formers, or those that actually aid crystallization, is shown quantitatively as follows:

$MgSO_4$	causes to crystallize		10	times its weight of sugar.		
$MgCl_2$	"	"	17	"	"	"
$Ca(NO_3)^2$	"	"	4	"	"	"
$CaCl_2$	"	"	7.5	"	"	"

The results of Marschall are at variance with those of other chemists in regard to individual salts; and though the conditions are scarcely such as obtain in the sugar manufacture, they are of considerable value in showing the general tendency of the various salts in this relation.

La Grange,* working after the general method of Mar-

* *La Sucrerie Indigène*, x. 259.

schall, considers, of all salts, the chlorides to have the least melassigenic effect, sodium chloride having scarcely any, the sulphates and carbonates coming next, and the alkaline nitrates most of all. The following table shows his results for several salts with their saline coefficient, or the proportion of their own weight of sugar they can render uncrystallizable:

	Yield in cane-sugar per 100 K.	Coeff.		Yield in sugar per 100 K.	Coeff.
Normal syrup.	54 K.	0	K_2CO_3	47 K.	3.50
NaCl	54 K.	0	KNO_3...	43 K.	5.50
KCl..........	48 K.	3.00	$NaNO_3$.......	41 K.	6.50
$CaCl_2$........	53 K.	.56	PO_4Na_3	44 K.	5.00
Na_2SO_4	50 K.	2.00			
K_2SO_4........	47 K.	3.50			
Na_2CO_3	47 K.	3.50			

Champion and Pellet* give the saline coefficient for a mixture of—

(1) 2½ grammes potassium nitrate and 1½ grammes potassium chloride as.................... .77

(2) Of the organic bodies separated by subacetate of lead and sulphuretted hydrogen as...... 1.42

(3) Invert-sugar of the optically inactive kind occurring in commercial products............ .56

These results have reference to beet-juice or molasses.

For *cane-molasses* the coefficient of potassium chloride is............................ .9 to 1.0

Organic substances separated as above.......... .86

Invert-sugar.................................. .56

The authors assume that potassium chloride is 25 per

* *La Sucrerie Indigène*, xii. 210, 223, 257.

cent. of the total weight of cane-molasses ash, and that the balance has no melassigenic action.

Vivien and Maumenè consider that chlorides in general, and especially chloride of sodium, have no melassigenic action. See also Grobert (*Journ. d. Fabr. Sucre*, xx. No. 5; *Zeit. f. Rübenz.*, xxix. 806).

Various Reactions of Cane-Sugar.—*Oxalate*, *citrate*, *carbonate*, and *basic phosphate of calcium* are less soluble in a sugar solution than in pure water. According to Sostman, sugar-water dissolves calcium sulphate in proportionally greater quantity as the solution is more concentrated and as the temperature is elevated. The solution, when boiled, gives up a portion of the salt. According to Bouchardat, *nascent hydrogen* converts cane-sugar into mannite, dulcite, or alcohols such as ethylic, isopropylic, and hexylic. *Sulphydrate of ammonia*, heated in a sealed tube to 150° with cane-sugar, gives a sulphuretted ethereal oil. *Fluoride of boron* is not absorbed by sugar in the cold, but on heating it is taken up and the sugar blackened. *Tetrachloride of carbon* heated to 100° with sugar is gradually colored brown and black, while dextrose is not altered. *Oil of sesame*, mixed with its volume of hydrochloric acid of commerce and raised to the boiling-point with a solution of cane or invert sugar produces a rose-color, even if only $\frac{1}{20000}$ part of the sugar is present (Vidan). In the presence of cane-sugar a certain number of metallic salts are not precipitated, or only imperfectly, by ammonia.

Parasaccharose.—This body, according to Jodin, is produced by the fermentation of a cane-sugar solution containing ammonic phosphate, together with another sugar isomeric with dextrose. It has the same composition as cane-sugar, crystalline, very soluble in water and insoluble in alcohol of 90 per cent. Sp. rotatory power, 108°, varying a little with fluctuations of temperature. Its action on alkaline solution of oxide of copper is about half that of dextrose. It is not altered by heating with sulphuric acid.

Inactive Cane-Sugar is a substance produced by the combination of parasaccharose and levulose, and is formed when solutions containing in certain proportions cane-sugar, phosphate of sodium, and sulphate of ammonium are allowed free access to the air. It is uncrystallizable, does not reduce alkaline copper solution, and is inactive to polarized light. Dilute acids transform it to a levo-rotatory sugar having $[a]j = -69$, which reduces copper solution (Jodin*).

* *Bull. Soc. Chim.*, i. 366.

CHAPTER III.

Dextrose, Levulose, and Invert-Sugar.

DEXTROSE $C_6H_{12}O_6$.

Glucose—Grape-Sugar—Right-handed Sugar—Sucre de Rasin, Fr.—Krumelzucker, Traubenzucker, Gr.; and, according to its origin—Fruit-Sugar—Honey-Sugar—Starch-Sugar—Diabetic Sugar—Rag-Sugar—Harnzucker, Gr.

DEXTROSE was first noticed by Lowitz and Proust, prepared from starch by Kirckhoff, and from linen by Braconnot. Its combinations with bases have been chiefly studied by Peligot,* and with organic acids by Berthelot.† Dubrunfaut has also added much to our knowledge of the chemical history of dextrose.‡

Dextrose occurs widely distributed in the vegetable kingdom in sweet fruits and grape-juice, associated often with cane-sugar and levulose; with the latter often in such a proportion as to constitute invert-sugar. It is also found in honey and numerous cereals. Many animal liquids and tissues contain dextrose, as the liver, blood, chyle, the yolk and white of eggs. Diabetic urine often holds dextrose to the amount of eight to ten per cent., as does the healthy secretion in small quantity (Bence Jones).

---

* *Ann. Chim. Phys.*, [2] lxvii. 136. † *Ibid.*, [3] liv. 74; lx. 95.

‡ *Ibid.*, [3] liii. 73; xxi. 169, 178; *Compt. Rend.*, xxiii. 38; xxv. 308; xxix. 51; xxxii. 249; xlii. 228, 739.

Formation.—By the transformation of carbohydrates with the assumption of water—as

$$\underset{\text{Starch.}}{3C_6H_{10}O_5} + H_2O = \underset{\text{Dextrose.}}{C_6H_{12}O_6} + \underset{\text{Dextrin.}}{2C_6H_{10}O_5}.$$

The above change takes place when starch is boiled with dilute acids. If the acid is allowed to act for some time, the dextrin first formed is converted into dextrose. Starch is also converted into dextrose by long boiling with water, and continued contact with gluten, saliva, and nitrogenous matters.

Glycogen and *lichenin* are changed to dextrose by boiling with dilute acids. When cellulose is treated with oil of vitriol, strong hydrochloric acid, or concentrated solution of zinc chloride, diluted, and the solution thus obtained boiled, dextrose is formed. *Tunicin*, under the same circumstances, is converted into dextrose. Maltose, melezitose, trehalose, and mycose give rise to dextrose by boiling with dilute acids.

Kosman * has found that grape-sugar or dextrose may be formed from glycerin and cellulose in the presence of air, water, and metallic iron, according to the reaction:

$$2C_3H_8O_3 + O_2 = C_6H_{12}O_6 + 2H_2O.$$

Glucosides, by boiling with dilute acids, produce dextrose and a non-saccharine body, by assumption of water. The transformation may be illustrated by one case:

$$\underset{\text{Amygdalin.}}{C_{20}H_{27}NO_{11}} + 2H_2O = \underset{\text{Dextrose.}}{2C_6H_{12}O_6} + \underset{\text{Bitter-almond oil.}}{C_7H_6O} + \underset{\text{Hydrocyanic acid.}}{CHN}.$$

Preparation.—From Starch by the Action of Dilute Acids.—One part of starch is boiled with four parts

* *Bull. Soc. Chim.*, xxviii. 246.

of water, and oil of vitriol in the quantity of $\frac{1}{10}$ to $\frac{1}{100}$ of the starch, the mixture stirred and kept at its original volume by the addition of water, until it is no longer precipitated by alcohol. From six to thirty-six hours of boiling are required, according to the amount of sulphuric acid present. The free acid is then neutralized by chalk, the liquid evaporated to 20° B., allowed to stand to deposit impurities, or is clarified, if necessary, with white of egg, filtered through animal charcoal, and the filtrate evaporated to a thick syrup from which the sugar separates after a few weeks. This product should be recrystallized.

PREPARATION OF PURE DEXTROSE (F. Soxhlet*).—One kilogramme of refined white cane-sugar is mixed with three litres of 90 per cent. alcohol and 120 c.c. pure, strong hydrochloric acid, and heated at 45° C. for two hours to invert. After ten days' standing crystals of dextrose begin to form, and in thirty-six hours dextrose is largely thrown down in crystals and powder. The deposit is washed with 90 per cent. and absolute alcohol, being finally recrystallized from the purest methylic alcohol (.810 sp. gr. for a quick crystallization, and .820 sp. gr. for a slower one).

PREPARATION FROM DIABETIC URINE.—Add excess of sodium chloride, when the glucosate is formed, which easily crystallizes out from a concentrated solution. The crystals are purified by washing with a saturated solution of salt, and finally with alcohol. The purified crystals are then dissolved in water, treated with sulphate of silver, filtered, and the mixture of dextrose and sodium sulphate evaporated to dryness on a water-bath. Strong alcohol

---

* See also Schwarz, Dingler, ccv. 427; Muspratt-Kerl, *Handbuch*, vi. 2078; Neubauer, *Zeit. Rübenz.*, 1876, 782.

dissolves out the pure dextrose from the residue, leaving the sulphate.

Properties.—From alcohol of 95 per cent. anhydrous dextrose is deposited in microscopic, well-defined needles, which melt at 146° C. to a colorless, transparent mass. Anhydrous dextrose is obtained as a white powder by heating hydrated dextrose to 60° C. in a stream of air. Crystallized dextrose dissolves at first quickly in water, but as the solution becomes more concentrated the action becomes much slower, so that several days are necessary for water to take up the full amount it is capable of dissolving. The concentrated syrup has not the elasticity or ropiness of cane-sugar syrup, and is disposed to be stringy when drawn out. 100 parts of water at 15° C. dissolve 81.68 parts of anhydrous dextrose and 97.85 parts of hydrated dextrose. The saturated solution contains 44.96 per cent. anhydrous dextrose; sp. gr. 1.206. According to Anthon, by dissolving hard crystallized dextrose in warm water a solution of density 1.221 at 17½° may be obtained. Dextrose seems to dissolve more readily when foreign matters are present. The sp. gr. of dextrose solution differs somewhat from that of cane-sugar containing the same amount of substance. The following table is given by Pohl*:

* *Wien. Akad. Ber.*, ii. 664.

Per cent. sugar.	Density of solution.		Difference in density.
	Cane-Sugar.	Grape-Sugar.	
2	1.0080	1.0072	— 8
5	1.0201	1.0200	— 1
7	1.0281	1.0275	— 6
10	1.0405	1.0406	+ 1
12	1.0487	1.0480	— 7
15	1.0616	1.0616	+ 0
17	1.0704	1.0693	— 11
20	1.0838	1.0831	— 7
22	1.0929	1.0909	— 20
25	1.1068	1.1021	— 47

Dextrose is soluble in aqueous alcohol in varying proportions, less easily, however, than cane-sugar. Anthon * gives the following solubilities: 1 part of dextrose requires for solution 50 parts alcohol of .837 sp. gr.

11.37 " " " .880 "
5.21 " " " .910 "
2.07 " " " .950 "

Melted dextrose deliquesces, and then solidifies to a crystalline hydrate. On evaporation of a solution the thick syrup does not solidify until sufficient water is absorbed to form a hydrate. Crystals separating from an alcoholic solution are hydrous or anhydrous, according to the strength of the alcohol; insoluble in ether.

Specific Rotatory Power.—For the anhydrous $[a]j =$ 52.5°, Clerget; 53.2°, Dubrunfaut; 55.1°, Pasteur; 56°, Berthelot; 57°, Schmidt; 57.4°, Béchamp; 57.7°, Jodin. Tollens, as the result of later investigations,† gives the following general formulas:

For the hydrate when the solution contains 8 to 91 per cent. of active substance:

---

* Dingler, *Polyt. Journal*, clv. 386. † *Ber. Chem. Gesell.*, ix. 1531.

$$[\alpha] D = 47.925 + .015534\, p + .0003883\, p^2;$$

for the anhydrous p = 7 to 83.

$$[\alpha] D = 52.718 + .017087\, p + .0004271\, p^2:$$

p = the percentage of sugar dissolved.

Hoppe-Seyler * has obtained for dextrose extracted from urine a value of

$$[\alpha] D = 56.4° \text{ (anhydrous)}; \; c = 14 \text{ to } 29 \text{ grm.}$$

See also Hesse.†

A freshly-prepared solution of hydrated dextrose, or dehydrated dextrose prepared without fusion, shows a rotatory power equal to nearly twice the above, but which gradually sinks to the normal, and then remains constant; by heating, the excessive rotation may be destroyed at once. Dubrunfaut has called the sugar having this property *birotatory dextrose;* and the phenomenon itself, *birotation.*

Composition.—

	Centesimally.	In equivalents.
Carbon................	40.00	72
Hydrogen..............	6.67	12
Oxygen................	53.33	96
	100.00	180

Decompositions—Heat.—When dextrose dried at 100° C. is heated to 170° C., it gives off two molecules of water and is converted into glucosan (Gelis‡); at 210° C. to 220° C. it swells, gives off more water, and yields caramel. The products formed at a high temperature are similar to those

* *Fres. Zeitschrift*, xiv. 305.

† *Ann. der Chem.*, clxxvi. 102.

‡ *Compt. Rend.*, li. 331.

obtained from cane-sugar under the same circumstances, but are somewhat more fusible, more easily soluble in water, and less soluble in alcohol. The products of the electrolytic decomposition of dextrose are hydrogen, oxygen, carbonic oxide, carbonic acid, the solution containing acetic and formic acids, and aldehyde.

When heated with *chromic acid*, or *peroxide of manganese and sulphuric acid*, formic acid is produced. Potassium dichromate warmed with aqueous dextrose does not alter it, but the presence of the latter prevents the reaction with cane-sugar. *Nascent hydrogen* converts dextrose into mannite. *Fuming nitric acid* forms nitro-dextrose; with ordinary or moderately dilute acid in the heat, *saccharic* and *oxalic* acids are formed, but no *tartaric* acid. Warmed with one molecule of *acid carbonate of potassium* and one of *iodine*, iodoform is produced (Millon). *Bromine* heated in a sealed tube with dextrose yields hydrobromic acid and a humus-like product; *chlorine* has a similar action. Bichloride of tin acts in the same manner as upon cane-sugar.

Cold *concentrated sulphuric acid*, when triturated with dextrose, dissolves it without coloration, forming a conjugated compound: *glucoso-sulphuric acid* $C_{24}H_{40}SO_{27}$. On heating charring takes place. Boiled with dilute acids, ulmin and ulmic acid are formed. According to Gautier,* when gaseous hydrochloric acid is passed into a cooled alcoholic solution of dextrose, an isomer of cane-sugar is formed having a bitter taste, soluble in water and alcohol, and which reduces cupric oxide in alkaline solution.

Action of Alkalies.—Gaseous *ammonia* is readily absorbed by dextrose when heated to 100° to 110°, water containing ammonic carbonate distilling off and a nitrogenous

---

* *Ber. Chem. Gesell.*, vii. 1549.

residue being left. Dextrose is decomposed by long contact with *alkalies*, *alkaline earths*, and some metallic oxides, forming glucic acid (Peligot); when heated with potash lye, the mixture becomes dark brown, smells of caramel, and contains *glucic acid* $C_{12}H_{18}O_9$ and *melassic acid* $C_{12}H_{10}O_5$. E. Feltz * gives the products of decomposition by heating with caustic alkalies, as saccharic, glucic, and apoglucic acids, of which the first two reduce the oxide of copper in alkaline solution in small quantity. *Alkaline carbonates* and *aqueous ammonia* produce the same effect as potash lye.

Lime distilled with a thick syrup of dextrose yields an oil from which *phorone* and *metacetone* may be obtained. *Baryta-water* boiled with dextrose, out of contact with the air, furnishes a solution which at first is yellow, but becomes dark on ebullition, and then contains glucate of baryta and another baryta salt from which aceto-formic acid $C_3H_4O_3\ 2H_2O$ may be obtained by distillation with dilute sulphuric acid.

Various Reactions.—Solution of *carbonate of soda* heated with dextrose and basic nitrate of bismuth produces a black-brown liquid and a grayish-brown precipitate; † this may serve as a qualitative reaction in the presence of cane-sugar and in urine. *Oxide of lead* heated to 110° with dextrose converts it into melassic acid in whole or part. *Ferric sulphate* and chloride are reduced to the ferrous salts on boiling with aqueous dextrose. Mixed with *nitrate of cobalt* and a small quantity of fused caustic potash, the solution remains clear on boiling, or, if very concentrated, deposits a light-brown precipitate; the presence of

---

* *Sucrerie Indigène*, vii. 165. † Boettger, *Journ. Pk. Chemie*, li. 431.

glucose prevents the appearance of the violet-blue precipitate with cane-sugar in this reaction.

When *cupric sulphate* in solution is mixed with aqueous dextrose and potash lye, the cupric hydrate, which at first separates, dissolves with a deep blue color, and deposits cuprous oxide after some time in the cold, and immediately when heated; this reaction is sensitive to detect and distinguish 1-100000 part of dextrose in the presence of cane-sugar, starch, or gum; under favorable circumstances, by the reddish color which the liquid assumes without precipitation, 1-1000000 part of dextrose may be shown (Trommer *); compare Guibourt.† Carbonic acid is formed in this reaction, as is also formic when cane-sugar is in excess, together with a peculiar body, resembling humic acid, which remains in combination with the alkali.

If dextrose is mixed with *indigo* solution, and the liquid boiled, carbonate of sodium solution being dropped in at the same time, the liquid is decolorized by the reduction of the indigo. ‡ *Nitrate of silver* boiled with dextrose throws down metallic silver as a black precipitate. An aqueous solution of one part *ferricyanide of potassium*, mixed with a half part of caustic *potash* and heated to 60° to 80°, is decolorized when dextrose is added; invert-sugar behaves in the same way, but cane-sugar and dextrin prepared by roasting do not (see page 210). If oil of vitriol is gradually added to an aqueous solution of *ox-gall*, until the precipitate first formed is redissolved, the liquid assumes a violet-red color, similar to that of potassium permanganate, on the addition of cane-sugar, dextrose, or starch (Pettenkofer); according to Van Brock, the extractive matter of

* *Ann. Pharm.*, xxxix. 361. † *Neues Journ. Pharm.*, xii. 263.

‡ Mulder. Neubauer, *Zeit. f. Anal. Chemie*, i. 377.

healthy urine and the reagents themselves produce this coloration in the absence of sugar. Dextrose absorbs oxygen readily, and reduces the salts of gold, silver, and bismuth to the metal, being oxidized to formic, oxalic, and tartronic acids and aldehyde (page 185).

COMBINATIONS.

With Water.—Dextrose forms two hydrates:

(*a*) HEMI-HYDRATED DEXTROSE, $2C_6H_{12}O_6 H_2O$ (Anthon's hard crystallized glucose). Prepared by a secret process.*

(*b*) MONO-HYDRATED DEXTROSE $C_6H_{12}O_6 H_2O$. Obtained in white, granular, hemispherical or cauliflower shaped masses with occasional shining faces. It loses some water at 65°–70°, and in a vacuum at 90°–100° it becomes anhydrous.

With Bases.—*Alkalies*, *alkaline earths*, and *plumbic oxide* form compounds with dextrose which are more easily decomposed than similar compounds of cane-sugar. Aqueous dextrose takes up a large quantity of the oxides of *barium*, *calcium*, and *strontium*, forming yellow solutions precipitated by alcohol, which, even when protected from the air, become darker, and are decomposed by standing or when exposed to heat; their taste is bitter and slightly alkaline; when evaporated *in vacuo*, a transparent, brittle mass remains containing unaltered dextrose.

With Potassium and Sodium Oxides.—Soluble in hot alcohol, the former crystallizable. According to Honig and Rosenfeld, on adding sodium ethylate to an alcoholic solution of dextrose, the compound $C_6H_{11}NaO_6$ is precipitated in white flocks.†

---

* *Chem. Centblatt*, 1859, 289. † *Dingler's Journal*, ccxxxvii. 146, 153.

With Barium Oxide.—(*a*) Obtained by precipitating a solution of dextrose in wood-spirit with baryta dissolved in aqueous wood-spirit; the precipitate is washed with the latter and dried *in vacuo* (Peligot). (*b*) Produced by adding alcoholic barium hydrate to excess of dextrose dissolved in alcohol, washing the precipitate with strong alcohol, and drying *in vacuo*. It is a nearly white, loose powder of caustic taste, and is easily soluble in water.

With Lime.* $C_6H_{10}CaO_6 + Aq$ (Peligot),
$2(C_6H_{12}O_6)CaO(H_2O)^2$ (Maumenè).

Prepared by adding alcohol to a freshly-made mixture of the sugar with calcic hydrate. Insoluble precipitate, difficult to dry.

With Plumbic Oxide.—Aqueous dextrose in the cold dissolves plumbic oxide, forming an insoluble basic compound which decomposes even below 100°. Aqueous dextrose gives no precipitate with neutral or basic acetate of lead, but gives one with the ammoniacal acetate.

(*a*) $C_6H_8Pb_2O_6$.
(*b*) $C_{12}H_{22}O_{11}$ 3PbO.

With Cupric Oxide.—Salkowski † describes a compound of cupric oxide which is formed as an insoluble precipitate, drying in the air to a blue-green powder partly soluble in alkali.

With Sodium Chloride.—The rotatory power of dextrose is not altered in the presence of sodium chloride.

(*a*) $C_6H_{12}O_6$ 2NaCl (nearly). Obtained by evaporating sodium chloride with diabetic urine (Staedeler).

(*b*) $2C_6H_{12}O_6$ 2NaCl H_2O. Also obtained by evaporating diabetic urine with sodium chloride.

* *Ann. Chim. Pharm.*, lxxxiii. 188. † *Zeits. f. Anal. Chemie*, xii. 98.

(*c*) $2C_6H_{12}O_6\ NaCl\ H_2O$. This is most well-defined of all the compounds of sodium chloride with dextrose. It crystallizes out when diabetic urine is concentrated; also from solutions containing one molecule or less of the salt to two molecules of the sugar. Dextrose from urine forms this body more easily than that from any other source. It consists of transparent, colorless, lustrous crystals, attaining a half-inch in length, belonging, according to Pasteur, to the right prismatic or rhombic system. Rotatory power $[\alpha]j = 47.14°$, corresponding to the unaltered rotation of the dextrose contained (Pasteur); permanent in the air; loses water when heated.

With Sodium Bromide.* $NaBr\ 2(C_6H_{12}O_6)$.

With Sodium Iodide.—A very unstable compound.

With Organic Acids.—Dextrose combines with the organic acids tartaric, stearic, benzoic, butyric, and acetic, forming amorphous solid or oily masses, soluble in alcohol and ether, but slightly soluble in water.†

Borax behaves with dextrose similarly to what it does with cane-sugar. By the action of *chloracetyl* on dextrose there is produced a body having the composition $C_6H_7(C_2H_3O)_4O_5Cl$. It has a specific rotatory power of 140°, reduces cupric oxide in alkaline solution, and is partially volatile. Dextrose prevents the precipitation of ferric chloride by alkalies.

ADDITIONAL QUALITATIVE TESTS FOR DEXTROSE.

Barfoed's Test.—Heat the solution with a neutral or acid solution of cupric acetate, when a precipitate is pro-

* Stenhouse (*Chem. Centb.*, 1864, 64).

† Berthelot (*Jahresb. der Chem.*, 1855, 157, 507, 678).

duced. Dextrin, milk, or cane-sugar do not act. .01 per cent. may be thus detected.*

Schmitt's Test.—A solution containing dextrose mixed with neutral acetate of lead and ammonia, gives a whitish cloud, which on warming settles to a red precipitate of lead sucrate. Cane-sugar, under the same conditions, gives a white precipitate only. A small trace of dextrose in the presence of much cane-sugar colors the deposit. Mannite acts as cane-sugar.

Mazzara's Test.—Hydrated sesquioxide of nickel, when heated with dextrose, invert-sugar, and many other organic bodies in the presence of caustic potash, is reduced to the green protoxide.† See also E. Pollacci.‡

Picric Acid Test (Braun §).—Dextrose reduces picric acid $C_6H_3(NO_2)_3O$ to picraminic acid $C_6H_3(NO_2)_2NH_2O$, the yellow color changing to a deep red. To execute the test the grape-sugar solution is heated with excess of caustic soda to 90° C., and one or two drops of a solution of picric acid added (containing 1 part acid to 250 parts water), and the whole heated to boiling.

Lindo's Test.‖—When the yellow crystalline compound obtained by the action of nitric acid on brucine is rendered alkaline by caustic alkali solution, and dextrose is added, the yellow color changes to an intense blue.

Paradextrose.—This substance is produced with parasaccharose by spontaneous fermentation of a cane-sugar solution (see page 72). It is isomeric with dextrose, forming a hydrate with one molecule of water. It loses its water at 100° and decomposes. Paradextrose does not

* *Fres. Zeitschrift*, xii. 27.

† *Gazzetta Chim. Ital.*, 1878, ii. and iii.

‡ *Ibid.*

§ *Fres. Zeitschrift*, iv. 187.

‖ *Chem. News*, xxxviii. 145.

reduce alkaline solution of tartrate of copper so strongly as dextrose, but by boiling with dilute acids its reducing power is increased. Sp. rotatory power, about +40°.

LEVULOSE * $C_6H_{12}O_6$.

Honigzucker, Linksfruchtzucker, Schleimzucker, Gr.—Chylariose, Fr.—Fruit-Sugar—Left-handed Glucose.

Levulose exists in the invert-sugar of honey and many fruits, though its isolated occurrence has not been demonstrated with certainty. Some fruits furnish a left rotatory juice, which renders it probable that levulose often exists in fruits in greater proportion than that necessary to form invert-sugar. The total sugar in such cases doubtless consists of a mixture of cane-sugar, dextrose, and levulose, or dextrose and levulose, the latter always predominating.

Formation.—(1) In the inversion of cane-sugar by water, dilute acids, yeast, or a peculiar ferment present in fruits; (2) by boiling levolusan with water or dilute acids. The sugar produced by the continued heating of *inulin* with acids is levulose, according to Dubrunfaut.

Preparation.—Add a little hydrochloric acid to a solution of cane-sugar, and heat to 60° C. When about twelve per cent. of invert-sugar is present, cool to —5° C. and add milk of lime, when the temperature will rise to 2° C. Submit the mixture to pressure to eliminate the liquid lime compound of dextrose, and to the levulosate of lime remaining add some water, and again express. Repeat this operation until the liquid running off has no longer a

---

* Bouchardat, *Compt. Rend.*, xxv. 274; Dubrunfaut and others, *ibid.*, xxix. 51, xl. 201, xlii. 803; *Ann. Chim. Phys.*, [3] xxi. 169; *Journ. Pk. Chem.*, lxix. 438, 208, xlii. 418.

dextro-rotation. The lime compound is then decomposed with oxalic acid. Finally, the solution of the pure levulose is submitted to cold by means of snow and hydrochloric acid, whereby the water freezes out, and the syrupy levulose remaining is further dried in a vacuum (Girard).

Properties.—Levulose is a colorless, uncrystallizable syrup or an amorphous mass. After heating to 100° its composition corresponds to the formula $C_6H_{12}O_6$. It is rather sweeter than cane-sugar, and is purgative. Optical rotatory power varying with the temperature, and much affected by presence of caustic lime.

$$[\alpha]_j = -\ 53^\circ \text{ at } 90^\circ \text{ C.}$$
$$-\ 79.5^\circ \text{ at } 52^\circ \text{ C.}$$
$$-\ 106^\circ \text{ at } 14^\circ \text{ C.}$$

(Dubrunfaut).

Neubauer makes the rotation at 14° to be − 100°.

Decompositions.—Levulose, on being heated to 170° C., yields a product analogous to glucosan, but more easily decomposed—probably levolusan $C_6H_{10}O_5$ (page 79). In contact with yeast it undergoes vinous fermentation without previous change. When sodium amalgam is added to an aqueous solution of invert-sugar, evolution of hydrogen ceases as soon as the liquid has become alkaline, heat is given off, and when the action is complete the solution is found to contain mannite (Linneman*). Levulose heated with dilute sulphuric acid forms *levulinic acid* (Grote and Tollens); it reduces cupric oxide in alkaline solution in the same proportion as dextrose. Chlorine, according to Hlasiwetz and Haberman,† with silver oxide, acting on levulose, forms not *gluconic* but *glycollic acid.*

* *Ann. Pharm.*, cxxiii. 136; *Ber. Chem. Gesell.*, ix. 1465; *Ann. Chim. Phys.*, [5] x. 559.

† *Ber. Chem. Gesell.*, iii. 486.

The products of the action of alkalies on levulose are the same as those obtained in the case of dextrose. They are the more complex in proportion as the air has access.* The decompositions and reactions of levulose in general are much the same as those of dextrose.

Combination with Lime.—Levulose forms with lime a basic compound analogous to that of dextrose, which absorbs oxygen from the air and decomposes. Another compound, consisting of sparingly soluble microscopic needles, containing three molecules of base to one of sugar, is decomposed by water when exposed to the light and air (Dubrunfaut †). Peligot gives the formula $C_6H_7O_7 3CaO$. Levulose is more soluble in alcohol than dextrose. A combination of sodium with levulose appears to exist, according to Honig and Rosenfeld,‡ of the formula $C_6H_{11}NaO_6$.

INVERT-SUGAR.

This is a mixture in equal equivalents of dextrose and levulose, produced by the action of heat, diastase, acids, salts, or other agents on cane-sugar and some of its isomers. It is an uncrystallizable syrup of sweeter taste than cane-sugar. The sp. rotatory power varies with the temperature : $[a]j$ =

at 14°	52°	90°
− 26.65°	− 13.33°	0° (Dubrunfaut §).

According to Tuchscmid,‖ 87.2° is the temperature of in-

* Peligot, *Compt. Rend.*, No. 4, 1880.

† *Ibid.*, lxix. 1366.

‡ *Ber. Chem. Gesell.*, xii. 45.

§ *Compt. Rend.*, xlii. 901.

‖ *Journ. Pk. Chemie*, [2] ii. 235.

activity. The latter gives the general formula when $c =$ 17.2 grm. in 100 c.c., as $[a]D_t = -(27.9 - .32t)$, when $t =$ the temperature. Alcohol lessens the left rotation of invert-sugar, especially in the heat.* Probably the invert-sugar in most commercial saccharine products is optically inactive at any temperature (page 173). In the inversion of cane-sugar by acids, for every 100° of the original dextro-rotation there is produced an inverse rotation of −38° at 15° C., and −44° at 0° C.

Chancel † finds that a contraction in volume takes place when a solution of cane-sugar is inverted, and hence invert-sugar solutions of the same percentage are heavier than those of cane-sugar.

COMPARATIVE DENSITIES OF CANE AND INVERT SUGAR SOLUTIONS.

Per cent. of sugar.	Density.		Per cent. of sugar.	Density.	
	Cane-sugar.	Invert-sugar.		Cane-sugar.	Invert-sugar.
2	1.0080	1.0082	15	1.0630	1.0634
5	1.0203	1.0206	17½	1.0718	1.0722
7	1.0286	1.0290	20	1.0854	1.0856
10	1.0413	1.0417	22	1.0946	1.0947
12	1.0499	1.0503	25	1.1086	1.1086

General references on invert-sugar: Bouchardat, *Compt. Rend.*, xxv. 274; Dubrunfaut, *N. Ann. Chim. Phys.*, xxi. 169; *Compt. Rend.*, xxix. 51, *ibid.*, xlii. 901, 803; Lippman, *Scheibler's Neue Zeit.*, iv. 304.

* Jodin, *Compt. Rend.*, lviii. 613.

† *Compt. Rend.*, lxxiv. 356.

CHAPTER IV.

LACTOSE OR MILK-SUGAR $C_{12}H_{22}O_{11}$.

Lactin—Sucre de Lait, Fr.—Milchzucker, Gr.

LACTOSE, an isomer of cane-sugar, is found in the milk of the mammalia. The only instance where it has been met with in a plant is in the *Achras sapota*, which furnishes lactose and another fermentable sugar in about equal proportions (Bouchardat, fils *).

Milk-sugar is prepared by heating milk with an acid or rennet, separating the curd, filtering through animal charcoal, if necessary, and evaporating to the crystallizing-point. It occurs in commerce generally as elongated, crystalline masses containing one equivalent of water.

The Specific Rotatory Power is to the right, and varies much with the concentration of the solution. It has been found by Hesse † to be for a solution containing

2 grm. of the hydrate to 100 c.c. $[a]D = 53.63°$.
3 " " " $= 53.16°$.
5 " " " $= 52.90°$.

The following is the general formula, wherein c = the number of grammes in 100 c.c. $c = 2$ to 12 ; temp. 15° C.

$$[a]D = 54.54 - .557c + .05475c^2 - .001774c^3.$$

---

* *Compt. Rend.*, Aug. 14, 1871. † *Ann. Chem. Pharm.*, clxxvi. 100.

Schmoeger* gives $[\alpha]D = 52.53°$ for a 30 per cent. solution at 20° C.

The freshly-prepared solution shows birotation. Alkalies alter the rotatory power.

Hydrated Lactose (crystallized lactose). $C_{12}H_{22}O_{11}$ H_2O. —Crystallizes in hard white or transparent four-sided hemihedral prisms belonging to the right prismatic system. Loses its water at 140° to 145°. Sp. gr., 1.543 at 13.9° C. Permanent in the air up to 100°.

Composition.—

	Anhydrous.	Hydrous.
Carbon................	42.11	40.00
Hydrogen..............	6.43	6.66
Oxygen................	51.46	53.34
	100.00	100.00

Solubilities.—Milk-sugar is slightly hygroscopic, and soluble in five or six parts of cold and 2.5 parts of boiling water. The saturated solution produced by contact with excess of the sugar has a density of 1.055, and contains 14.55 per cent. of the crystallized substance. This solution, when left to evaporation, deposits crystals as soon as the density reaches 1.063. Lactose dissolves readily in distilled vinegar, and crystallizes from it again unaltered. It is insoluble in absolute alcohol and ether (Dubrunfaut †).

Action of Heat.—At 150° C. lactose acquires a yellow color; at 160° C. gives off the smell of caramel and loses slightly in weight; at 175°, or above, it is partially converted, with loss of weight, into *lacto-caramel* and a sub-

* *Ber. Chem. Gesell.*, xvi. 1922.

† *Jahresber. der Chemie*, 1856, 643.

stance insoluble in water which melts at 203.5° C. (Lieben.*) Crystallized lactose, when carefully heated, gives off 12 per cent. of water, and solidifies on cooling to a crystalline mass, which on solution regains its water (Berzelius).

By dry distillation lactose yields carbonic acid, combustible gases, acetic acid, empyreumatic oil, and charcoal. Heated in the open air it swells up, becomes brown and tenacious, gives off the odor of burnt sugar, and leaves a large quantity of coal. By roasting, gum and saccharic acid are produced. In aqueous solution the sugar is decomposed when heated in a sealed tube to 100°–130°.

Sulphuric Acid (concentrated) chars lactose at 100°. Heated with the diluted acid, the optical rotatory power is increased three-tenths, *gallactose* being formed (lactose of Pasteur), and a partially dextro-rotatory, non-fermentable substance which is crystallizable (Dubrunfaut). According to Fudakowsky,† two sugars are produced in the reaction, both fermentable, soluble in water, dextro-gyrate, but differing by their solubility in alcohol. The sp. rotatory power, after warming and long standing, of the two are respectively

$$[a]\,D = 92.83°,$$
$$[a]\,D = 62.63°.$$

Both are birotatory.‡

Nitric Acid diluted, heated with milk-sugar, gives mucic, saccharic, tartaric, oxalic, and carbonic acids. The production of mucic acid in this way is particularly characteristic of lactose. Nitric acid may act on lactose in

* *Jahresber. der Chemie*, 1856, 643. † *Zeits. Anal. Chem.*, [2] iii. 32.

‡ See also Meissl, *Journ. Pk. Chem.*, xxii. 100.

two ways: 1. The greater part of the sugar may be converted into mucic acid, which then undergoes further decomposition, yielding tartaric acid. 2. A small portion of the sugar is changed, as in the case with sulphuric acid, into gallactose, the ultimate product of the reaction being tartaric acid. Strong nitric acid forms an explosive nitro-substitution compound.

Concentrated *hydrochloric acid* turns lactose brown, while *glacial phosphoric acid* forms a red color, but does not carbonize it; oxidized by *potassium chlorate* and *iodic acid.* Distilled with *potassium dichromate* and sulphuric acid, *aldehyde* is formed.

Action of Alkalies.—Milk-sugar absorbs 12.40 per cent. of *ammonia gas.* By action of *caustic potash* a compound is obtained from which acids separate the lactose unaltered. Triturated with potassium hydrate and water, a brown liquid containing acetic acid is obtained.

A solution containing three molecules of free alkali to one of *cupric oxide*, with tartaric acid or an alkaline tartrate, yields, when heated with milk-sugar, a precipitate of cuprous oxide.

Ritthausen* has obtained from milk, by the action of cupric sulphate and potassium hydrate, a carbohydrate which, after boiling with acids, reduces the alkaline solution of tartrate of copper. A. W. Blyth describes two new copper-reducing bodies from milk corresponding to the formulas CH_3O_6 and $C_3H_2O_4$, and considers them to be glucosides.

Lactose forms more or less well-defined compounds with potassium, sodium, calcium, barium, and lead oxides.†

---

* *Journ. Prak. Chemie, neue folge*, xv. 848.

† Honig and Rosenfeld, *Ber. Chem. Gesell.*, xii. 47.

There are two lime compounds—one soluble and containing the same number of molecules of base and sugar, and the other insoluble and basic. Schutzenberger,* by action of acetic anhydride on milk-sugar, obtained *octacetylated lactose*

$$C_{12}H_{14}(C_2H_3O)_8O_{11}\ [a]j = 31°,$$

and *quadriacetylated lactose*

$$C_{12}H_{18}(C_2H_3O)_4O_{11}\ [a]j = 50.1°.$$

See also Herzfeld.†

Lactose does not unite with *sodium chloride* in definite proportion.

Fermentation.—Milk-sugar ferments at 30° with yeast, but more slowly than grape-sugar or dextrose, yielding alcohol and carbonic acid. Milk ferments also spontaneously without the addition of yeast, producing alcohol. A solution of milk-sugar in contact with putrid caseine or gluten gives alcohol and lactic acid, the milk-sugar being not previously converted into gallactose. Less alcohol is obtained if the acid is neutralized as fast as formed.

Erythrozyme, a substance obtained from madder, causes milk-sugar to ferment, giving rise to carbonic, formic, acetic, and succinic acids, hydrogen, and alcohol.‡

---

* *Ann. Chem. Pharm.*, clx. 91. † Scheibler's *Neue Zeit.*, iii. 155.

‡ Schunck, *Journ. Pk. Chem.*, lxiii. 22.

CHAPTER V.

DETERMINATION OF SPECIFIC GRAVITY.

THE specific gravity or density of solids and liquids is a ratio expressing their weight relative to an equal volume of water at a standard temperature; this temperature is generally that of water at its greatest density, 4° C., though 15° C. is sometimes adopted. The specific gravity of water is called 1, and that of all other solid and liquid bodies consists of multiples of this, whether whole numbers or fractions; thus, the specific gravity of alcohol is .7938, and that of gold is 19.3—that is, a volume of alcohol or gold, respectively, weighs .7938 and 19.3 times as much as an equal bulk of water at 4° C.

There are three principal methods of determining the density of solids and liquids—viz.: 1. BY THE HYDROSTATIC BALANCE; 2. BY THE SPECIFIC-GRAVITY FLASK; and 3. BY THE AREOMETER. All of these are the same in principle, as they consist in ascertaining, directly or indirectly, the weight of a body in air, and that of an equal bulk of water.

The Hydrostatic Balance.—The use of this piece of apparatus depends upon the following physical law, first enunciated by Archimedes: *A solid body immersed in a liquid loses a part of its weight equal to the weight of the displaced liquid.* Hence, if we weigh a solid on an ordinary balance, first in air, and then in water by suspending it with a fine hair or silk thread from the scale-pan, the difference between the two weighings will represent the weight of a volume of water equal to that of the solid; by dividing the weight in air by that of the bulk of water displaced, the specific gravity is obtained.

Mohr has devised a form of the hydrostatic balance whereby the determination of the density of liquids may be made with rapidity and accuracy; the principle of the apparatus is easily derived from the Archimedean theorem:

Fig. 4.

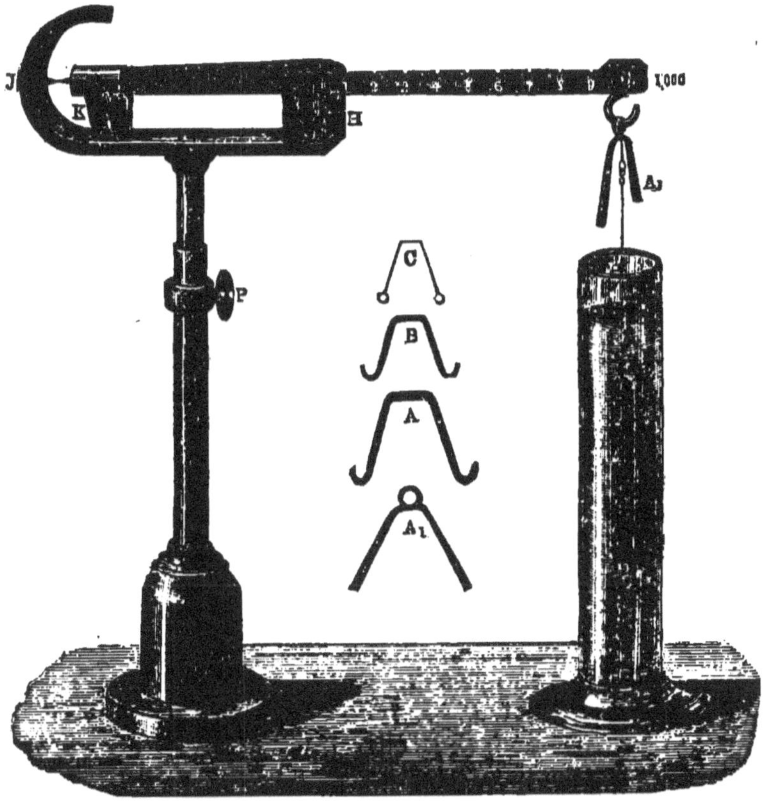

It is evident that if a solid is weighed while suspended in a liquid, that the decrease in weight from that of the same body in air, or volume displaced, must be proportional to the density of the liquid. The apparatus (Fig. 4) consists of a beam which is in equilibrium in air when the sinker—a glass cylinder enclosing a thermometer and hung to the extremity of the arm by a fine platinum wire—is attached. It is necessary to have the balance perfectly horizontal, and for this purpose a small levelling-table may be used; there is an elevating-screw, P, by which the beam may be raised or lowered to suit the requirements of the operation. The depth to which the sinker should descend below the level of the liquid under examination will not vary much from that shown in the figure. The weights consist of a series of decimal riders, of which A A_1 (Fig. 4) are equal to each other, and likewise equal to the weight of the volume of distilled water displaced by the sinker at 15° C.; B is one-tenth of A, and C is one-hundredth. A, B, and C are hung on the graduated beam. When the sinker is immersed in distilled water at 15° C., and the rider A_1 is on the end of the beam, as in Fig. 4, the balance is in equilibrium and corresponds to the density of 1.000. For liquids heavier than water the other riders are placed on the beam, A_1 still hanging on its extremity, until equilibrium is restored; the riders when thus placed have following values:

$$A_1 = 1.000$$
$$A = .100$$
$$B = .010$$
$$C = .001$$

For liquids lighter than water A_1 is taken off and the balance restored with the other riders. Fig. 5 shows examples of the readings with different densities. The sinker and platinum wire, after use, should be cleaned and dried with care.

By the Specific-Gravity Flask.—This is the simplest and at the same time one of the most accurate methods of taking the specific gravity of solid and liquid bodies. The apparatus is a tared flask holding to a mark on the neck a determinate weight of distilled water at 15° C. or 4° C. The liquid to be examined is brought to the required temperature and filled into the flask up to the mark, and the

Fig. 5.

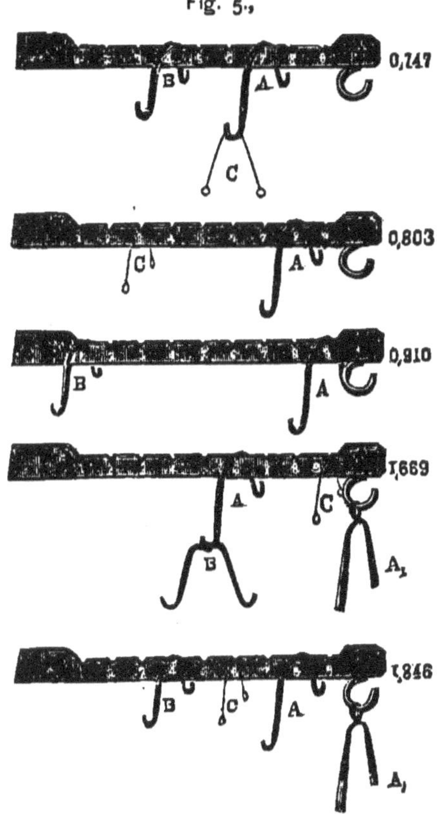

whole weighed; the last weight, after subtracting that of the flask, is that of a bulk of the liquid equal to the volume of water whose weight is known; the density is then obtained on dividing the former by the latter.

The 100 c.c. flask is a very convenient arrangement for

taking specific gravities, the weight of 100 c.c. of any liquid, divided by 100, being its specific gravity.

AREOMETRY.

The Areometer (*Arädometer, Senkwage, Gr.; Aréomètre, Fr.*)—The areometer consists of a closed tube expanded below into a bulb the lower part of which is loaded to maintain the instrument in an upright position when floating.

According to the laws of hydrostatics, a body immersed in a fluid is buoyed up with a force exactly equal to the weight of the volume displaced; hence, if the body float, the weight of the bulk displaced is equal to that of the floating body; the weight of an areometer in air is, therefore, the same as that of the volume of liquid displaced by it when floating freely.

Areometers may be divided into two classes—viz., (1) *those having constant volume and variable weight*, and (2) *those of variable volume and constant weight.* Hydrometers of the first class, on being placed in the liquid to be tested, sink to a fixed mark on the stem by means of weights added; from these weights the volume displaced is calculated. Nicholson's and Fahrenheit's hydrometers are of this kind. Those of the second class are provided with a scale on the stem, and the instruments, when used, are allowed to sink in the liquid until they float in equilibrium, the point at which the surface of the liquid cuts the scale indicating either directly the specific gravity, or, in the case of areometers with arbitrary scale, merely a degree which does not show directly the density. In regard to areometers with variable volume, it may be said that if a floating body of constant weight be immersed in different fluids, their densities will be in inverse ratio to the volumes displaced; the less dense the greater the displacement, and *vice versa.* Suppose we float a hydrometer weighing 50 grammes; then the volume of liquid displaced by it will weigh exactly 50 grammes. Now, with fluids of various densities the 50 grammes will correspond to volumes inversely as the density. In the case of water the volume displaced would be 50 c.c. = 50 grammes, which, divided by the weight of the areometer, gives $\frac{50}{50} = 1.000$ as the specific gravity; 50 grammes syrup of specific gravity 1.261 would occupy a volume of 39.6 c.c. $\left(\frac{50}{1.261}\right)$. Accordingly, the division of the scale shown by areometers corresponds to volume displaced, and either shows directly the specific gravity, or a formula may be obtained by which the indications of the arbitrary scales may be reduced to specific gravities.

The scales of areometers of variable volume are *even* or *uneven*—the former include the majority of hydrometers in ordinary use; the latter are those in which the graduations read specific gravities, and are called *densimeters.* Even scale hydrometers for use in the arts, and, indeed, for scientific purposes, have some advantages over those reading specific gravities. The scale of the latter, being expressed in decimal fractions, are more difficult to remember and record; for example, it is easier to remember that a solution is 25° Baumé than that its specific gravity is 1.2173. The densimeter is, furthermore, much more difficult to construct correctly, and consequently more costly. The scales of all areometers should, however, be based on fixed and invariable data, so that the specific gravity corresponding to any degree may be calculated. Such data, and tables based on them, are given in another part of this work.

As areometers are used for two different classes of liquids —those denser than water, which increase in value with the density; and those less dense, which contain more of the valuable constituent the lower the specific gravity, water at a standard temperature is the natural zero-point for hydrometer scales; for fluids heavier than water the degrees will read downwards, while for those lighter the readings will be in reverse order; in either case the number of divisions of the scale from zero will increase in proportion to the amount of valuable constituent present in the solution examined. Hence for arbitrary scales the reading is natural, easily comprehended and remembered.

The form of the part of the hydrometer below the surface of the liquid may vary, but it should be symmetrical with the axis, or otherwise the instrument would not float perfectly upright, but would lean; the lower part is always more or less expanded, so that the stem may not be of inordinate length. The greater the volume of the bulb proportional to that of the stem, the greater will be the sensitiveness of the instrument; that is, a small difference in density or displacement will correspond to a large space on the stem. It has on this account been found useful to graduate hydrometers for special purposes in which the scale extends only over a limited field of density; in this way each degree may occupy a larger space and may be divided into fractions. Examples of these are Baumé's hydrometers called acidometers, salinometers, alcoholometers, saccharometers, etc., made for special purposes in the arts.

Hydrometers are made of glass, or metal, as brass, silver, or German silver. Glass is preferable for most purposes from its cheapness and the ease with which it is worked; besides which air bubbles adhere less to glass than metallic surfaces, thus lessening one of the greatest sources of error inherent with the use of hydrometers, especially when dense solutions are operated upon. Another advantage of glass is the impossibility of indenting the surface, which is a source of error to which metallic areometers are peculiarly liable. Glass is not, however, suitable as a material for very sensitive areometers, because the extreme smallness of bore it is necessary to give the stem would make it too fragile.

The areometers of variable volume in common use essentially differ in the manner in which the scale is divided. The following are those which will be described in this work:

I. When the scale indicates volumes displaced—*Gay Lussac's volumeter* (even scale).

II. When the scales indicate directly specific gravities—*the densimeter* (uneven scale).

III. When the indications of the scale are arbitrary—*Baumé's hydrometer* (even scale).

IV. When the scale indicates percentages of substance in solution—*Balling's saccharometer* (even scale).

GAY LUSSAC'S VOLUMETER.

This areometer is of the ordinary form, and the scale shows directly the volume displaced of the liquid tested with it, in comparison with that of water with the same instrument. Thus, if it is floated in a solution and stands at 40° on the scale, this indicates for the same weight, the volume of water being 100, that of the solution would be 40; whence the density may be readily calculated. The

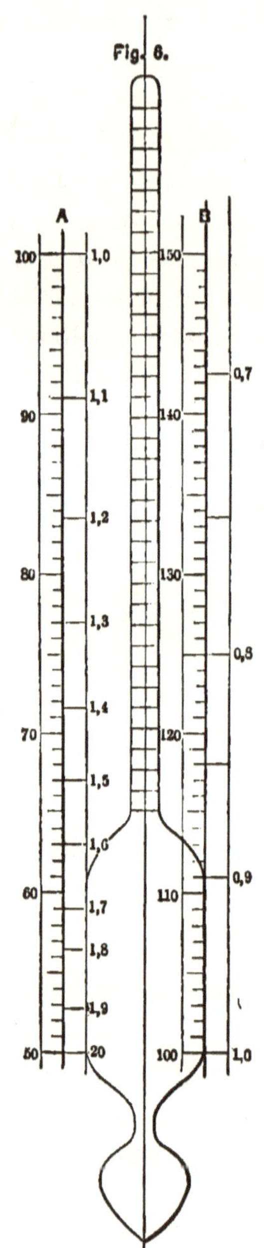

point to which the instrument sinks in water is marked 100 on the scale. Above and below this, divisions are made of such a kind that the volume of the stem comprised between two successive degrees is $\frac{1}{100}$ of the total volume below the 100° mark. As the volumeter is more exact the larger the divisions of the scale, it is advisable to have it made in two spindles, one for liquids heavier than water, with the 100° point at the upper part of the scale, Fig. 6, A; and another for liquids lighter than water, with the 100° point near the bottom, Fig. 6, B.

In order to obtain the specific gravity of a liquid it is simply necessary to divide the volume displaced in water 100, by the number on the scale to which the apparatus sinks. The same rule applies for liquids lighter than water. Thus, if the volumeter marks 120, we have $\frac{100}{120}$, or .833. In the figure the scales to the right and left are those of the volumeter with the corresponding specific gravities opposite. It is seen that equal differences in volume correspond to unequal differences in density.

The volumeter has a great advantage over the densimeter in that its scale is even. It has all the advantage of an areometer with arbitrary scale, its degrees being whole numbers (though reading inversely for liquids heavier than water), and yet, by a very simple calculation, the indications may be converted into specific gravities.

The table below gives the correspondence of the volumeter and specific gravities:

Degree of volumeter.	Density.	Degree of volumeter.	Density.
50.	2.000	90.90	1.1000
52.63	1.900	100.00	1.0000
55.55	1.800	105.26	.9500
58.82	1.700	111.11	.9000
62.50	1.600	117.64	.8500
66.66	1.500	125.00	.8000
71.43	1.400	133.33	.7500
76.92	1.300	142.85	.7000
83.33	1.200		

THE DENSIMETER.

This instrument reads directly, without calculation, the specific gravity of the liquid in which it floats. The scale is so made that the point to which the hydrometer sinks in distilled water at standard temperature is marked 1.000, and the graduation for liquids lighter than water is carried above this point, and for liquids heavier than water in the reverse direction. The finer hydrometers of this kind have the scale divided between two or more spindles, so that the increased length of stem gives room for a more accurately-divided scale. When the densimeter consists only of the hydrominor and hydromajor spindles, the 1.000 point is placed with the first at the bottom of the areometer, and at the top with the second.

Ventzke's Areometer is a densimeter with a bulb enormously enlarged compared with the stem, which is very short and thin, in order that the instrument may have great sensitiveness. The middle of the stem, is marked with density of 1.100, and divisions showing small differences in density are carried above and below this point. The areometer is used in Ventzke's process for determining sugar by the optical saccharimeter, and also for estimating the water in sugars and syrups from their density when in solution (see page 147).

BAUMÉ'S HYDROMETER.

Baumé's hydrometer is generally made of glass, of the ordinary form, and loaded with shot or mercury. The scale may be either engraved on the stem, or of paper enclosed within it, as is the form of the cheaper kinds. There are two entirely distinct Baumé hydrometers, graduated on different principles, the divisions of their scales not being directly convertible into each other. They are the *hydromajor* and the *hydrominor* spindles. For the hydromajor instrument (*pèse sel*, *pèse acide*) the point marked 15° on the scale was fixed by Baumé at the place on the scale where it sinks in a solution of common salt made by dissolving fifteen parts by weight in eighty-five parts of water. The space between this and zero was divided into fifteen equal parts, and divisions of the same size were continued below 15° to the bulb.

In the hydrominor spindle (*pèse spirit*) the point on the stem to which it sinks in water is marked 10, while the zero is where it stands in a solution of ten parts common salt in ninety parts of water. The density of this solution is 1.0847. The distance between 0° and 10° is divided into

10 equal parts, and this division is extended to the rest of the scale.

Standard of Graduation.—There has always been some uncertainty about the standard proposed by Baumé for fixing the points for the graduation of his areometers. He himself prescribes the use of pure and dry salt, which would yield a solution of sp. gr. 1.109 for the hydromajor spindle. Other authorities direct the use of common salt which contains varying quantities of moisture and from two to eight per cent. of other impurities, varying with the quality. Hence it is very evident that hydrometers graduated by the two methods will have scales not comparable. If, however, the directions of Baumé are rigidly adhered to, and a solution of chemically-pure salt, of sp. gr. 1.109 at 15°, is used, there could be no better or more unvarying standard. A new method for graduating these hydrometers was introduced by Gay Lussac, by which the zero-point corresponds to distilled water at 4° C., and the degree at which they stand in pure monohydrated sulphuric acid is made 66° at 15° C., the intermediate part of the scale between these two points being divided in 66 equal parts. At present all Baumé's hydrometers are graduated on Gay Lussac's plan, except that both of the fixed points are generally taken at the temperature of 15° C.; the difference between this way of graduating and the unmodified one of Gay Lussac is too small to be taken into consideration, unless in very exceptional circumstances. The chief practical objection to this method is that the oil of vitriol of commerce used by makers of these instruments varies in density considerably, as coming from different sources; also, the densities of the pure acid, as given by various authorities, differ sufficiently to cause a serious error in the graduation based on these data. These differences are probably owing for the most part to the varying temperature at which the specific gravity was taken. That given by Gay Lussac is 1.8427 at 15° C., and is entirely reliable. It will be seen that the areometers graduated with oil of vitriol, without regard to temperature, or a strict determination of the density of the graduating liquid so that it may be the same as the figure given above, will show a notable error; but if regard is paid to the necessary conditions of the operation, and these conditions are the same in all cases, the hydrometers agree very closely with each other, and their readings can always be converted into specific gravities by appropriate formulas.

A good hydrometer has a stem of the same calibre throughout, and the scale equally divided. The accuracy in these respects may be readily determined with a pair of calipers and compasses.

Reduction of Scale to Specific Gravity.– Though the scale of the Baumé instrument is arbitrary, yet the specific gravity corresponding to any degree may be calculated. Tables of these equivalents, in the case of hydrometers for liquids heavier than water, met with in the books, show great discrepancies, for the reason that some are calculated by the following formula given by Francœur:

$$P = \frac{152}{152 - d} \quad (1)$$

in which P = the density; d = the degree Baumé. This is the correct formula when the graduation is effected by the original method of Baumé with a solution of salt. When Gay Lussac's method is used with sulphuric acid of sp. gr. 1.8427 at 15° C., the formula becomes

$$P = \frac{144.3}{144.3 - d} \quad (2)$$

Tables calculated after (2) are the only ones practically useful, as the instruments are no longer graduated with salt solution.

The formula for the hydrominor hydrometer is

$$P = \frac{146}{136 + d} \quad (3)$$

The following table given by Bourgougnon * shows the specific gravities corresponding to degrees Baumé for liquids heavier than water, at 15° C., calculated according to formula (2) :

Deg. B.	Sp. Gr.	Deg. B.	Sp. Gr.	Deg. B.	Sp. Gr.	Deg. B.	Sp. Gr.
0	1.0000	19	1.1516	38	1.3574	57	1.6527
1	1.0069	20	1.1608	39	1.3703	58	1.6719
2	1.0140	21	1.1702	40	1.3834	59	1.6915
3	1.0212	22	1.1798	41	1.3968	60	1.7115
4	1.0285	23	1.1895	42	1.4104	61	1.7321
5	1.0358	24	1.1994	43	1.4244	62	1.7531
6	1.0433	25	1.2095	44	1.4386	63	1.7748
7	1.0509	26	1.2197	45	1.4530	64	1.7968
8	1.0586	27	1.2301	46	1.4678	65	1.8194
9	1.0665	28	1.2407	47	1.4829	66	1.8427
10	1.0744	29	1.2514	48	1.4983	67	1.8665
11	1.0825	30	1.2624	49	1.5140	68	1.8909
12	1.0906	31	1.2735	50	1.5301	69	1.9161
13	1.0989	32	1 2849	51	1.5465	70	1.9418
14	1.1074	33	1.2964	52	1.5632	71	1.9683
15	1.1159	34	1.3081	53	1.5802	72	1.9955
16	1.1246	35	1.3201	54	1.5978	73	2.0235
17	1.1335	36	1.3323	55	1.6157	74	2.0523
18	1.1424	37	1.3447	56	1.6340	75	2.0819

* *Proc. Am. Chem. Soc.*, vol. i., No. 5, p. 55.

The table for liquids lighter than water is calculated by formula (3):

Deg. B.	Sp. Gr.	Deg. B.	Sp. Gr.	Deg. B.	Sp. Gr.	Deg. B.	Sp. Gr.
10	1.000	23	.918	36	.849	49	.789
11	.993	24	.913	37	.844	50	.785
12	.986	25	.907	38	.839	51	.781
13	.980	26	.901	39	.834	52	.777
14	.973	27	.896	40	.830	53	.773
15	.967	28	.890	41	.825	54	.768
16	.960	29	.885	42	.820	55	.764
17	.954	30	.880	43	.816	56	.760
18	.948	31	.874	44	.811	57	.757
19	.942	32	.869	45	.807	58	.753
20	.936	33	.864	46	.802	59	.749
21	.930	34	.859	47	.798	60	.745
22	.924	35	.854	48	.794		

Correction for Temperature.—As the areometer, especially in the sugar industry, is often used at temperatures above the ordinary, it is desirable to obtain a correction that will serve to reduce the readings to the degree of heat at which the instruments are graduated. A correction amply accurate enough for ordinary purposes, or, indeed, to any purpose to which this hydrometer is itself suited, may be deduced from the results of the following experiments given in foot-note.*

When the temperature is above 15° C. or 62° F., the product of the number of degrees in excess, multiplied by .0471 or .0265, is *added* to the hydrometer reading; when it is below the standard temperature, it is *subtracted;* or the correction of $\frac{1}{10}$ degree Baumé for each difference of two degrees Centigrade may be used.

BALLING'S AREOMETER.

The readings of this areometer, sometimes called *Bal-*

* Molasses of two densities and a strong syrup of cane-sugar were heated

ling's saccharometer, indicate directly the percentage of pure sugar or solid matter dissolved. Thus, if it is floated in a solution of pure cane-sugar and sinks to 30°, the liquid contains thirty parts of sugar and seventy parts of water. If the solution contains other matters besides pure sugar, the readings show percentages of dissolved matter or impure sugar. The form is that of a bulb loaded with mercury, carrying a long stem on which is the scale. For accurate instruments the whole scale is not on one spindle, but there are three, one embracing the scale from zero to 30°, the second from 25° to 60°, and the third from 55° to 90°. The degrees are divided into halves or fifths to allow of more exact results.

Balling, an Austrian chemist, originated this hydrometer, and made careful determinations of the specific gravities of sugar solutions corresponding to various percentages of sugar dissolved, as did also Nieman and Gerlach.

---

successively to different temperatures, and the readings of the hydrometer carefully taken and averaged.

1. Molasses stood :

At 11° C. — 34.3° B. 40° C. — 32.9° B. 56° C. — 32.2° B. 79° C. — 31.2° B.	Average for 1° C. = a difference of .0456° B.

2. Molasses :

At 13° C. — 15.6° B. 39° C. — 14.7° B. 69° C. — 13.3° B. 87° C. — 12.1° B.	Average 1° C. = a difference of .0473 B.

3. A syrup of cane-sugar stood :

At 12° C. — 33.25° B. 48° C. — 31.20° B. 77° C. — 30.10° B.	Average 1° C. = a difference of .0484° B.

Average of the three estimations :

1° C. = .0471° B.
1° F. = .0265° B.

Later, Brix has recalculated Balling's table, making some corrections, and now the instrument is made according to the results of Brix, and is generally known as the Brix saccharometer. The terms *Brix's* or *Balling's areometer* or *saccharometer* will be used indifferently in this work.

Error due to Impurities.—It is evident that when Balling's saccharometer is used on impure sugar solutions, the indications will be incorrect in proportion as the specific gravity of the impurities differs from that of cane-sugar, and that the error will also be proportionally greater as the impurities exist in larger quantity in comparison with the cane-sugar. The following table shows the density of some of the leading impurities contained in cane or beet juice:

	20 per cent. solution.	25 per cent. solution.
Cane-sugar	1.0833	1.10607
Grape "	1.0831	1.10210
Calcium acetate	1.0874	1.1130
" nitrate	1.1736	1.2220
Sodium sulphate	1.0807	1.1017
" nitrate	1.1418	1.1832
Potassium nitrate	1.1359	......

Another table, given by Frese,* shows the same class of facts. The solutions each contain one per cent. of substance:

* Frese, *Beitrage der Zuckerfabrikation.*

	Sp. Gr.	Per cent. Balling.
Potassium carbonate	1.0086	2.15
" hydrate	1.0088	2.20
" " neut. with PO_4H_3	1.0156	3.90
" nitrate	1.0062	1.55
" hydrate neut. with citric acid	1.0110	2.75
Sodium carbonate	1.0084	2.10
" sulphate	1.0044	1.10
" chloride	1.0070	1.75
" phosphate	1.0040	1.00
" oxalate	1.0036	.90
" citrate	1.0036	.90

With grape-sugar the density for strong solutions is sufficient to make a difference of about one per cent., while for some of the salts the error is enormously greater.

This source of inaccuracy will always prevent Balling's saccharometer from giving perfectly reliable results in solutions containing much impurity, though there can be no doubt of its great value for ordinary technical work, even on the lower products of the *fabricant* and refiner.

Correction for Temperature.—This correction is given in the following table, arranged from that of Stammer.* It is to be observed that when the temperature of the solution operated upon is lower than 17½° C. (63½° F.), the correction is to be subtracted from the reading of the areometer; when above 17½° it is to be added. If the saccharometer is graduated at 15° C. instead of 17½°, the difference made by using the table given below is too small to be considered in ordinary work:

* *Lehrbuch der Zuckerfabrikation;* Ergänzungband, page 60.

CORRECTION FOR THE READINGS OF BALLING'S SACCHAROMETER, ON ACCOUNT OF TEMPERATURE.

Temp.		Per cent. of sugar in solution.												
C.	F.	0	5	10	15	20	25	30	35	40	50	60	70	75
		To be subtracted from the degree read.												
0°	32°	.17	.30	.41	.52	.62	.72	.82	.92	.98	1.11	1.22	1.25	1.29
5	41	.23	.30	.37	.44	.52	.59	.65	.72	.75	.80	.88	.91	.94
10	51	.20	.26	.29	.33	.36	.39	.42	.45	.48	.50	.54	.58	.61
11	52	.18	.23	.26	.28	.31	.34	.36	.39	.41	.43	.47	.50	.53
12	53.6	.16	.20	.22	.24	.26	.29	.31	.33	.34	.36	.40	.42	.46
13	55.4	.14	.18	.19	.21	.22	.24	.26	.27	.28	.29	.33	.35	.39
14	57.0	.12	.15	.16	.17	.18	.19	.21	.22	.22	.23	.26	.28	.32
15	59.0	.09	.11	.12	.14	.14	.15	.16	.17	.16	.17	.19	.21	.25
16	61.0	.06	.07	.08	.09	.10	.10	.11	.12	.12	.12	.14	.16	.18
17	62.5	.02	.02	.03	.03	.03	.04	.04	.04	.04	.04	.05	.05	.06
		To be added to the degree read.												
18	64.4	.02	.03	.03	.03	.03	.03	.03	.03	.03	.03	.03	.03	.02
19	66.2	.06	.08	.08	.09	.09	.10	.10	.10	.10	.10	.10	.08	.06
20	68.0	.11	.14	.15	.17	.17	.18	.18	.18	.19	.19	.18	.15	.11
21	70.0	.16	.20	.22	.24	.24	.25	.25	.25	.26	.26	.25	.22	.18
22	71.6	.21	.26	.29	.31	.31	.32	.32	.32	.33	.34	.32	.29	.25
23	73.4	.27	.32	.35	.37	.38	.39	.39	.39	.40	.42	.39	.36	.33
24	75.0	.32	.38	.41	.43	.44	.46	.46	.47	.47	.50	.46	.43	.40
25	77.0	.37	.44	.47	.49	.51	.53	.54	.55	.55	.58	.54	.51	.48
26	79.0	.43	.50	.54	.56	.58	.60	.61	.62	.62	.66	.62	.58	.55
27	80.6	.49	.57	.61	.63	.65	.68	.68	.69	.70	.74	.70	.65	.62
28	82.4	.56	.64	.68	.70	.72	.76	.76	.78	.78	.82	.78	.72	.70
29	84.0	.63	.71	.75	.78	.79	.84	.84	.86	.86	.90	.86	.80	.78
30	86.0	.70	.78	.82	.87	.87	.92	.92	.94	.94	.98	.94	.88	.86
35	95.0	1.10	1.17	1.22	1.24	1.30	1.32	1.33	1.35	1.36	1.39	1.34	1.27	1.25
40	104.0	1.50	1.61	1.67	1.71	1.73	1.79	1.79	1.80	1.82	1.83	1.78	1.69	1.65
50	122	....	2.65	2.71	2.74	3.78	2.80	2.80	2.80	2.80	2.79	2.70	2.56	2.51
60	140	....	3.87	3.88	3.88	3.88	3.88	3.88	3.88	3.90	3.82	3.70	3.43	3.41
70	158	....	....	5.18	5.20	5.14	5.13	5.10	5.08	5.06	4.90	4.72	4.47	4.35
80	176	....	....	6.62	6.59	6.54	6.46	6.38	6.30	6.26	6.06	5.82	5.50	5.33

According to observations of Gerlach, the correction for temperature varies with the concentration of the solution and the range of temperature as shown in the table.

A very convenient form of the Balling saccharometer is to have the thermometer forming part of the areometer, with its stem enclosed within that of the latter. The thermometer is graduated into degrees Centigrade on one side of the stem, and into the corresponding corrections for some of the more common temperatures on the other; so that not only is the taking of the temperature as a separate operation dispensed with, but also the trouble of consulting the table. In this way the corrected degree Balling may be obtained by two readings and a simple mental operation.

Vivien's Saccharometer.—This areometer has two scales, one showing the number of kilos. of sugar in the hectolitre of sugar solution, and the other the corresponding specific gravities. It consists of three separate spindles, the first having a range from 1 to 1.025 sp. gr., the second from 1.025 to 1.05, and the third from 1.05 to 1.075. The instrument is intended especially for beet-juice or other thin saccharine liquids.

The following table gives the percentages of sugar, or degree Balling, of sugar solutions, with the corresponding densities and degrees Baumé. It was calculated by Mategczek, Scheibler, and Stammer.*

---

* *Zeitschrift für Zuckerindustrie des Deutschen Reiches*, xv. 583, xx. 269; Stammer, *Zuckerfabrikation*, 28 *et seq.*

Table showing the Relation of Percentages, Specific Gravities, and Degrees Baumé in Cane-Sugar Solutions.

Per cent. of Sugar.	Specific Gravity.	Degree Baumé.	Per cent. of Sugar.	Specific Gravity.	Degree Baumé.	Per cent. of Sugar.	Specific Gravity.	Degree Baumé.	Per cent. of Sugar.	Specific Gravity.	Degree Baumé.
0.0	1.0000	0.0	7.3	1.0290	4.1	14.6	1.0596	8.1	21.9	1.0918	12.1
.1	1.0003	0.06	.4	1.0294	4.1	.7	1.0600	8.15	22.0	1.0923	12.2
.2	1.0007	0.11	.5	1.0298	4.2	.8	1.0604	8.2	.1	1.0927	12.2
.3	1.0011	0.17	.6	1.0302	4.2	.9	1.0609	8.3	.2	1.0932	12.3
.4	1.0015	0.22	.7	1.0306	4.3	15.0	1.0613	8.3	.3	1.0936	12.3
.5	1.0019	0.28	.8	1.0310	4.3	.1	1.0617	8.4	.4	1.0941	12.4
.6	1.0023	0.33	.9	1.0314	4.4	.2	1.0621	8.4	.5	1.0945	12.4
.7	1.0027	0.39	8.0	1.0318	4.4	.3	1.0626	8.5	.6	1.0950	12.5
.8	1.0031	0.44	.1	1.0322	4.5	.4	1.0630	8.5	.7	1.0954	12.55
.9	1.0034	0.5	.2	1.0327	4.55	.5	1.0634	8.6	.8	1.0959	12.6
1.0	1.0038	0.55	.3	1.0331	4.6	.6	1.0639	8.65	.9	1.0964	12.7
.1	1.0042	0.6	.4	1.0335	4.7	.7	1.0643	8.7	23.0	1.0968	12.7
.2	1.0046	0.7	.5	1.0339	4.7	.8	1.0647	8.8	.1	1.0973	12.8
.3	1.0050	0.7	.6	1.0343	4.8	.9	1.0652	8.8	.2	1.0977	12.8
.4	1.0054	0.8	.7	1.0347	4.8	16.0	1.0656	8.9	.3	1.0982	12.9
.5	1.0058	0.8	.8	1.0351	4.9	.1	1.0660	8.9	.4	1.0986	12.9
.6	1.0062	0.9	.9	1.0355	4.9	.2	1.0665	9 0	.5	1.0991	13.0
.7	1.0066	0.9	9.0	1.0359	5.0	.3	1.0669	9.0	.6	1.0996	13.0
.8	1.0070	1.0	.1	1.0364	5.05	.4	1.0674	9.1	.7	1.1000	13.1
.9	1.0074	1.05	.2	1.0368	5.1	.5	1.0678	9.1	.8	1.1005	13.15
2.0	1.0077	1.1	.3	1.0372	5.2	.6	1.0682	9.2	.9	1.1009	13.2
.1	1.0081	1.2	.4	1.0376	5.2	.7	1.0687	9.25	24.0	1.1014	13.3
.2	1.0085	1.2	.5	1.0380	5.3	.8	1.0691	9.3	.1	1.1019	13.3
.3	1.0089	1.3	.6	1.0384	5.3	.9	1.0695	9.4	.2	1.1023	13.4
.4	1.0093	1.3	.7	1.0388	5.4	17.0	1.0700	9.4	.3	1.1028	13.4
.5	1.0097	1.4	.8	1.0393	5.4	.1	1.0704	9 5	.4	1.1032	13.5
.6	1.0101	1.4	.9	1.0397	5.5	.2	1.0709	9.5	.5	1.1037	13.5
.7	1.0105	1.5	10.0	1.0401	5.55	.3	1.0713	9.6	.6	1.1042	13.6
.8	1.0109	1.55	.1	1.04 5	5.6	.4	1.0717	9.6	.7	1.1046	13.6
.9	1.0113	1.6	.2	1.0409	5.7	.5	1.0722	9.7	.8	1.1051	13.7
3.0	1.0117	1.7	.3	1.0413	5.7	.6	1.0726	9.75	.9	1.1056	13.75
.1	1.0121	1.7	.4	1.0418	5.8	.7	1.0730	9.8	25.0	1.1060	13.8
.2	1.0125	1.8	.5	1.0422	5.8	.8	1.0735	9.9	.1	1.1065	13.9
.3	1.0129	1.8	.6	1.0426	5.9	.9	1.0739	9.9	.2	1.1070	13.9
.4	1.0133	1.9	.7	1.0430	5 9	18.0	1.0744	10.0	.3	1.1074	14.0
.5	1.0137	1.9	.8	1.0434	6.0	.1	1.0748	10.0	.4	1.1079	14.0
.6	1.0141	2.0	.9	1.0439	6.05	.2	1.0753	10.1	.5	1.1083	14.1
.7	1.0145	2.0	11.0	1.0443	6.1	.3	1.0757	10.1	.6	1.1088	14.1
.8	1.0149	2.1	.1	1.0447	6.2	.4	1.0761	10.2	.7	1.1093	14.2
.9	1.0153	2.2	.2	1.0451	6.2	.5	1.0766	10.2	.8	1.1097	14.2
4.0	1.0157	2.2	.3	1.0455	6.3	.6	1.0770	10.3	.9	1.1102	14.3
.1	1.0161	2.3	.4	1.0459	6.3	.7	1.0775	10.35	26.0	1.1107	14.35
.2	1.0165	2.3	.5	1.0464	6.4	.8	1.0779	10.4	.1	1.1111	14.4
.3	1.0169	2.4	.6	1.0468	6.4	.9	1.0783	10.5	.2	1.1116	14.5
.4	1.0173	2.4	.7	1.0472	6.5	19.0	1.0788	10.5	.3	1.1121	14.5
.5	1.0177	2.5	.8	1.0476	6.55	.1	1.0792	10.6	.4	1.1125	14.6
.6	1.0181	2.6	.9	1.0481	6.6	.2	1.0797	10.6	.5	1.1130	14.6
.7	1.0185	2.6	12.0	1.0485	6.7	.3	1.0801	10.7	.6	1.1135	14.7
.8	1.0189	2.7	.1	1.0489	6.7	.4	1.0805	10.7	.7	1.1140	14.7
.9	1.0193	2.7	.2	1.0493	6.8	.5	1.0810	10.8	.8	1.1144	14.8
5.0	1.0197	2.8	.3	1.0497	6.8	.6	1.0815	10.85	.9	1.1149	14.8
.1	1.0201	2.8	.4	1.0502	6.9	.7	1.0819	10.9	27.0	1.1154	14.9
.2	1.0205	2.9	.5	1.0506	6.9	.8	1.0824	11.0	.1	1.1158	14.9
.3	1.0209	2.9	.6	1.0510	7.0	.9	1.0828	11.0	.2	1.1163	15.0
.4	1.0213	3.0	.7	1.0514	7.05	20.0	1.0832	11.1	.3	1.1168	15.1
.5	1.0217	3.0	.8	1.0519	7.1	.1	1.0837	11.1	.4	1.1172	15.1
.6	1.0221	3.1	.9	1.0523	7.2	.2	1.0841	11.2	.5	1.1177	15.2
.7	1.0225	3.2	13.0	1.0527	7.2	.3	1.0846	11.2	.6	1.1182	15.2
.8	1.0229	3.2	.1	1.0531	7.3	.4	1.0850	11.3	.7	1.1187	15.3
.9	1.0233	3.3	.2	1.0536	7.3	.5	1.0855	11.3	.8	1.1191	15.3
6.0	1.0237	3.3	.3	1.0540	7.4	.6	1.0859	11.4	.9	1.1196	15.4
.1	1.0241	3.4	.4	1.0544	7.4	.7	1.0864	11.45	28.0	1.1201	15.4
.2	1.0245	3.4	.5	1.0548	7.5	.8	1.0868	11.5	.1	1.1206	15.5
.3	1.0249	3.5	.6	1.0553	7.5	.9	1.0873	11.6	.2	1.1210	15.55
.4	1.0253	3.6	.7	1.0557	7.6	21.0	1.0877	11.6	.3	1.1215	15.6
.5	1.0257	3.6	.8	1.0561	7.65	.1	1.0882	11.7	.4	1.1220	15.7
.6	1.0261	3.7	.9	1.0566	7.7	.2	1.0886	11.7	.5	1.1225	15.7
.7	1.0265	3.7	14.0	1.0570	7.8	.3	1.0891	11.8	.6	1.1229	15.8
.8	1.0269	3.8	.1	1.0574	7.8	.4	1.0895	11.8	.7	1.1234	15.8
.9	1.0273	3.8	.2	1.0578	7.9	.5	1.0900	11.9	.8	1.1239	15.9
7.0	1.0277	3.9	.3	1.0583	7.9	.6	1.0904	11.95	.9	1.1244	15.9
.1	1.0281	3.9	.4	1.0587	8.0	.7	1.0909	12.0	29.0	1.1248	16.0
.2	1.0286	4.0	.5	1.0591	8.0	.8	1.0914	12.05	.1	1.1253	16.0

Per cent. of Sugar.	Specific Gravity.	Degree Baume.	Per cent. of Sugar.	Specific Gravity.	Degree Baume.	Per cent. of Sugar.	Specific Gravity.	Degree Baume.	Per cent. of Sugar.	Specific Gravity.	Degree Baume.
29.2	1.1258	16.1	37.1	1.1646	20.35	45.0	1.2056	24.6	52.9	1.2489	28.7
.3	1.1263	16.1	.2	1.1651	20.4	.1	1.2061	24.6	53.0	1.2495	28.75
.4	1.1267	16.2	.3	1.1656	20.5	.2	1.2067	24.7	.1	1.2500	28.8
.5	1.1272	15.25	.4	1.1661	20.5	.3	1.2072	24.7	.2	1.2506	28.85
.6	1.1277	16.3	.5	1.1666	20.6	.4	1.2077	24.8	.3	1.2512	28.9
.7	1.1282	16.4	.6	1.1671	20.6	.5	1.2083	24.8	.4	1.2517	28.9
.8	1.1287	16.4	.7	1.1676	20.7	.6	1.2088	24.9	.5	1.2523	29.0
.9	1.1291	16.5	.8	1.1681	20.7	.7	1.2093	24.9	.6	1.2529	29.1
30.0	1.1296	16.5	.9	1.1686	20.8	.8	1.2099	25.0	.7	1.2534	29.1
.1	1.1301	16.6	38.0	1.1692	20.8	.9	1.2104	25.0	.8	1.2540	29.2
.2	1.1306	16.6	.1	1.1697	20.9	46.0	1.2110	25.1	.9	1.2546	29.2
.3	1.1311	16.7	.2	1.1702	20.9	.1	1.2115	25.1	54.0	1.2551	29.3
.4	1.1315	16.7	.3	1.1707	21.0	.2	1.2120	25.2	.1	1.2557	29.3
.5	1.1320	16.8	.4	1.1712	21.05	.3	1.2126	25.2	.2	1.2563	29.4
.6	1.1325	16.85	.5	1.1717	21.1	.4	1.2131	25.3	.3	1.2568	29.4
.7	1.1330	16.9	.6	1.1722	21.15	.5	1.2136	25.35	.4	1.2574	29.5
.8	1.1335	17.0	.7	1.1727	21.2	.6	1.2142	25.4	.5	1.2580	29.5
.9	1.1340	17.0	.8	1.1732	21.3	.7	1.2147	25.45	.6	1.2585	29.6
31.0	1.1344	17.1	.9	1.1737	21.3	.8	1.2153	25.5	.7	1.2591	29.6
.1	1.1349	17.1	39.0	1.1743	21.4	.9	1.2158	25.6	.8	1.2597	29.7
.2	1.1354	17.2	.1	1.1748	21.4	47.0	1.2163	25.6	.9	1.2602	29.7
.3	1.1359	17.2	.2	1.1753	21.5	.1	1.2169	25.7	55.0	1.2608	29.8
.4	1.1364	17.3	.3	1.1758	21.5	.2	1.2174	25.7	.1	1.2614	29.8
.5	1.1369	17.3	.4	1.1763	21.6	.3	1.2180	25.8	.2	1.2620	29.9
.6	1.1374	17.4	.5	1.1768	21.6	.4	1.2185	25.8	.3	1.2625	29.9
.7	1.1378	17.4	.6	1.1773	21.7	.5	1.2191	25.9	.4	1.2631	30.0
.8	1.1383	17.5	.7	1.1778	21.7	.6	1.2196	25.9	.5	1.2637	30.05
.9	1.1388	17.55	.8	1.1784	21.8	.7	1.2201	26.0	.6	1.2642	30.1
32.0	1.1393	17.6	.9	1.1789	21.85	.8	1.2207	26.0	.7	1.2648	30.15
.1	1.1398	17.7	40.0	1.1794	21.9	.9	1.2212	26.1	.8	1.2654	30.2
.2	1.1403	17.7	.1	1.1799	22.0	48.0	1.2218	26.1	.9	1.2660	30.25
.3	1.1408	17.8	.2	1.1804	22.0	.1	1.2223	26.2	56.0	1.2665	30.3
.4	1.1412	17.8	.3	1.1809	22.1	.2	1.2229	26.2	.1	1.2671	30.4
.5	1.1417	17.9	.4	1.1815	22.1	.3	1.2234	26.3	.2	1.2677	30.4
.6	1.1422	17.9	.5	1.1820	22.2	.4	1.2240	26.35	.3	1.2683	30.5
.7	1.1427	18.0	.6	1.1825	22.2	.5	1.2245	26.4	.4	1.2688	30.5
.8	1.1432	18.0	.7	1.1830	22.3	.6	1.2250	26.45	.5	1.2694	30.6
.9	1.1437	18.1	.8	1.1835	22.3	.7	1.2256	26.5	.6	1.2700	30.6
33.0	1.1442	18.15	.9	1.1840	22.4	.8	1.2261	26.6	.7	1.2706	30.7
.1	1.1447	18.2	41.0	1.1846	22.4	.9	1.2267	26.6	.8	1.2712	30.7
.2	1.1452	18.25	.1	1.1851	22.5	49.0	1.2272	26.7	.9	1.2717	30.8
.3	1.1457	18.3	.2	1.1856	22.5	.1	1.2278	26.7	57.0	1.2723	30.8
.4	1.1462	18.4	.3	1.1861	22.6	.2	1.2283	26.8	.1	1.2729	30.9
.5	1.1466	18.4	.4	1.1866	22.65	.3	1.2289	26.8	.2	1.2735	30.9
.6	1.1471	18.5	.5	1.1872	22.7	.4	1.2294	26.9	.3	1.2740	31.0
.7	1.1476	18.5	.6	1.1877	22.75	.5	1.2300	26.9	.4	1.2746	31.0
.8	1.1481	18.6	.7	1.1882	22.8	.6	1.2305	27.0	.5	1.2752	31.1
.9	1.1486	18.6	.8	1.1887	22.9	.7	1.2311	27.0	.6	1.2758	31.1
34.0	1.1491	18.7	.9	1.1892	22.9	.8	1.2316	27.1	.7	1.2764	31.2
.1	1.1496	18.7	42.0	1.1898	23.0	.9	1.2322	27.1	.8	1.2769	31.2
.2	1.1501	18.8	.1	1.1903	23.0	50.0	1.2327	27.2	.9	1.2775	31.3
.3	1.1506	18.85	.2	1.1908	23.1	.1	1.2333	27.2	58.0	1.2781	31.3
.4	1.1511	18.9	.3	1.1913	23.1	.2	1.2338	27.3	.1	1.2787	31.4
.5	1.1516	18.95	.4	1.1919	23.2	.3	1.2344	27.3	.2	1.2793	31.4
.6	1.1521	19.0	.5	1.1924	23.2	.4	1.2349	27.4	.3	1.2799	31.5
.7	1.1525	19.1	.6	1.1929	23.3	.5	1.2355	27.45	.4	1.2804	31.5
.8	1.1531	19.1	.7	1.1934	23.3	.6	1.2361	27.5	.5	1.2810	31.6
.9	1.1536	19.2	.8	1.1940	23.4	.7	1.2366	27.55	.6	1.2816	31.6
35.0	1.1541	19.2	.9	1.1945	23.45	.8	1.2372	27.6	.7	1.2822	31.7
.1	1.1546	19.3	43.0	1.1950	23.5	.9	1.2377	27.7	.8	1.2828	31.7
.2	1.1551	19.3	.1	1.1955	23.55	51.0	1.2383	27.7	.9	1.2834	31.8
.3	1.1556	19.4	.2	1.1961	23.6	.1	1.2388	27.8	59.0	1.2840	31.85
.4	1.1561	19.4	.3	1.1966	23.7	.2	1.2394	27.8	.1	1.2845	31.9
.5	1.1566	19.5	.4	1.1971	23.7	.3	1.2399	27.9	.2	1.2851	31.95
.6	1.1571	19.55	.5	1.1976	23.8	.4	1.2405	27.9	.3	1.2857	32.0
.7	1.1576	19.6	.6	1.1982	23.8	.5	1.2411	28.0	.4	1.2863	32.05
.8	1.1581	19.65	.7	1.1987	23.9	.6	1.2416	28.0	.5	1.2869	32.1
.9	1.1586	19.7	.8	1.1992	23.9	.7	1.2422	28.1	.6	1.2875	32.15
36.0	1.1591	19.8	.9	1.1998	24.0	.8	1.2427	28.1	.7	1.2881	32.2
.1	1.1596	19.8	44.0	1.2003	24.0	.9	1.2433	28.2	.8	1.2887	32.3
.2	1.1601	19.9	.1	1.2008	24.1	52.0	1.2439	28.2	.9	1.2893	32.3
.3	1.1606	19.9	.2	1.2013	24.1	.1	1.2444	28.3	60.0	1.2898	32.4
.4	1.1611	20.0	.3	1.2019	24.2	.2	1.2450	28.3	.1	1.2904	32.4
.5	1.1616	20.0	.4	1.2024	24.2	.3	1.2455	28.4	.2	1.2910	32.5
.6	1.1621	20.1	.5	1.2029	24.3	.4	1.2461	28.4	.3	1.2916	32.5
.7	1.1626	20.1	.6	1.2035	24.35	.5	1.2467	28.5	.4	1.2922	32.6
.8	1.1631	20.2	.7	1.2040	24.4	.6	1.2472	28.5	.5	1.2928	32.6
.9	1.1636	20.2	.8	1.2045	24.45	.7	1.2478	28.6	.6	1.2934	32.7
37.0	1.1641	20.3	.9	1.2051	24.5	.8	1.2483	28.65	.7	1.2940	32.7

Per cent. of Sugar.	Specific Gravity.	Degree Baumé.	Per cent. of Sugar.	Specific Gravity.	Degree Baumé.	Per cent. of Sugar.	Specific Gravity.	Degree Baumé.	Per cent. of Sugar.	Specific Gravity.	Degree Baumé.
60.8	1.2946	32.8	67.2	1.3334	36.0	73.5	1.3732	39.1	79.8	1.4145	42.2
.9	1.2952	32.8	.3	1.3340	36.0	.6	1.3738	39.1	.9	1.4152	42.2
61.0	1.2958	32.9	.4	1.3346	36.1	.7	1.3745	39.2	80.0	1.4158	42.2
.1	1.2964	32.9	.5	1.3352	36.1	.8	1.3751	39.2	.1	1.4165	42.3
.2	1.2970	33.0	.6	1.3359	36.2	.9	1.3757	39.3	.2	1.4172	42.3
.3	1.2975	33.0	.7	1.3365	36.2	74.0	1.3764	39.3	.3	1.4179	42.4
.4	1.2981	33.1	.8	1.3371	36.3	.1	1.3770	39.4	.4	1.4185	42.4
.5	1.2987	33.1	.9	1.3377	36.3	.2	1.3777	39.4	.5	1.4192	42.5
.6	1.2993	33.2	68.0	1.3384	36.4	.3	1.3783	39.5	.6	1.4199	42.5
.7	1.2999	33.2	.1	1.3390	36.4	.4	1.3790	39.5	.7	1.4205	42.6
.8	1.3005	33.3	.2	1.3396	36.5	.5	1.3796	39.6	.8	1.4212	42.6
.9	1.3011	33.3	.3	1.3402	36.5	.6	1.3803	39.6	.9	1.4219	42.7
62.0	1.3017	33.4	.4	1.3408	36.6	.7	1.3809	39.7	81.0	1.4226	42.7
.1	1.3023	33.4	.5	1.3415	36.6	.8	1.3816	39.7	.1	1.4232	42.8
.2	1.3029	33.5	.6	1.3421	36.7	.9	1.3822	39.8	.2	1.4239	42.8
.3	1.3035	33.5	.7	1.3427	36.7	75.0	1.3828	39.8	.3	1.4246	42.9
.4	1.3041	33.6	.8	1.3433	36.8	.1	1.3835	39.9	.4	1.4253	42.9
.5	1.3047	33.6	.9	1.3440	36.8	.2	1.3842	39.9	.5	1.4259	43.0
.6	1.3053	33.7	69.0	1.3446	36.9	.3	1.3848	40.0	.6	1.4265	43.0
.7	1.3059	33.7	.1	1.3452	36.9	.4	1.3855	40.0	.7	1.4273	43.1
.8	1.3065	33.8	.2	1.3458	37.0	.5	1.3861	40.1	.8	1.4280	43.1
.9	1.3071	33.8	.3	1.3465	37.0	.6	1.3868	40.1	.9	1.4287	43.2
63.0	1.3077	33.9	.4	1.3471	37.1	.7	1.3874	40.2	82.0	1.4293	43.2
.1	1.3083	33.9	.5	1.3477	37.1	.8	1.3880	40.2	.1	1.4300	43.3
.2	1.3089	34.0	.6	1.3484	37.2	.9	1.3887	40.3	.2	1.4307	43.3
.3	1.3095	34.0	.7	1.3490	37.2	76.0	1.3894	40.3	.3	1.4314	43.4
.4	1.3101	34.1	.8	1.3456	37.3	.1	1.3900	40.4	.4	1.4320	43.4
.5	1.3107	34.1	.9	1.3502	37.3	.2	1.3907	40.4	.5	1.4327	43.5
.6	1.3113	34.2	70.0	1.3509	37.4	.3	1.3913	40.5	.6	1.4334	43.5
.7	1.3119	34.2	.1	1.3515	37.4	.4	1.3920	40.5	.7	1.4341	43.5
.8	1.3126	34.3	.2	1.3521	37.5	.5	1.3926	40.6	.8	1.4348	43.6
.9	1.3132	34.3	.3	1.3528	37.5	.6	1.3933	40.6	.9	1.4354	43.6
64.0	1.3138	34.4	.4	1.3534	37.6	.7	1.3940	40.7	83.0	1.4361	43.7
.1	1.3144	34.4	.5	1.3540	37.6	.8	1.3946	40.7	.1	1.4368	43.7
.2	1.3150	34.5	.6	1.3546	37.7	.9	1.3953	40.8	.2	1.4375	43.8
.3	1.3155	34.5	.7	1.3553	37.7	77.0	1.3959	40.8	.3	1.4382	43.8
.4	1.3162	34.6	.8	1.3559	37.8	.1	1.3966	40.8	.4	1.4388	43.9
.5	1.3168	34.6	.9	1.3565	37.8	.2	1.3972	40.9	.5	1.4395	43.9
.6	1.3174	34.7	71.0	1.3572	37.9	.3	1.3979	41.0	.6	1.4402	44.0
.7	1.3180	34.7	.1	1.3578	37.9	.4	1.3986	41.0	.7	1.4409	44.0
.8	1.3186	34.8	.2	1.3585	38.0	.5	1.3992	41.0	.8	1.4416	44.1
.9	1.3192	34.8	.3	1.3591	38.0	.6	1.3999	41.1	.9	1.4423	44.1
65.0	1.3198	34.9	.4	1.3597	38.1	.7	1.4005	41.1	84.0	1.4430	44.2
.1	1.3205	34.95	.5	1.3604	38.1	.8	1.4012	41.2	.1	1.4437	44.2
.2	1.3211	35.0	.6	1.3610	38.2	.9	1.4019	41.2	.2	1.4443	44.3
.3	1.3217	35.05	.7	1.3616	38.2	78.0	1.4025	41.3	.3	1.4450	44.3
.4	1.3223	35.1	.8	1.3623	38.2	.1	1.4032	41.3	.4	1.4457	44.3
.5	1.3229	35.15	.9	1.3629	38.3	.2	1.4039	41.4	.5	1.4464	44.4
.6	1.3235	35.2	72.0	1.3635	38.3	.3	1.4045	41.4	.6	1.4471	44.4
.7	1.3241	35.25	.1	1.3642	38.4	.4	1.4052	41.5	.7	1.4478	44.5
.8	1.3247	35.3	.2	1.3648	38.4	.5	1.4058	41.5	.8	1.4485	44.5
.9	1.3253	35.35	.3	1.3655	38.5	.6	1.4065	41.6	.9	1.4492	44.6
66.0	1.3260	35.4	.4	1.3661	38.5	.7	1.4072	41.6	85.0	1.4498	44.6
.1	1.3266	35.4	.5	1.3667	38.6	.8	1.4078	41.7	.1	1.4505	44.7
.2	1.3272	35.5	.6	1.3674	38.6	.9	1.4085	41.7	.2	1.4512	44.7
.3	1.3278	35.5	.7	1.3680	38.7	79.0	1.4092	41.8	.3	1.4519	44.8
.4	1.3285	35.6	.8	1.3687	38.7	.1	1.4098	41.8	.4	1.4526	44.8
.5	1.3291	35.6	.9	1.3693	38.8	.2	1.4105	41.9	.5	1.4533	44.9
.6	1.3297	35.7	73.0	1.3699	38.8	.3	1.4112	41.9	.6	1.4540	44.9
.7	1.3303	35.7	.1	1.3705	38.9	.4	1.4119	42.0	.7	1.4547	45.0
.8	1.3309	35.8	.2	1.3712	38.9	.5	1.4125	42.0	.8	1.4554	45.0
.9	1.3315	35.8	.3	1.3719	39.0	.6	1.4132	42.1	.9	1.4561	45.1
67.0	1.3322	35.9	.4	1.3725	39.0	.7	1.4138	42.1	86.0	1.4568	45.1
.1	1.3327	35.0									

Another table is given, partially supplementary to the last and calculated by the same formulas, but taking in a wider range of densities, and having the degrees Baumé in the first column:

TABLE SHOWING RELATION BETWEEN DEGREES BAUMÉ, PERCENTAGES, AND SPECIFIC GRAVITIES OF CANE-SUGAR SOLUTIONS.

Degrees Baumé.	Degrees Brix.	Specific Gravity.	Degrees Baumé.	Degrees Brix.	Specific Gravity.	Degrees Baumé.	Degrees Brix.	Specific Gravity.	Degrees Baumé.	Degrees Brix.	Specific Gravity.
0.	.00	1.0000	13.	23.52	1.0992	26.	47.73	1.2203	39.	73.23	1.3714
0.5	.90	1.0035	13.5	24.43	1.1034	26.5	48.68	1.2255	39.5	74.25	1.3780
1.	1.80	1.0070	14.	25.35	1.1077	27.	49.63	1.2308	40.	75.27	1.3846
1.5	2.69	1.0105	14.5	26.27	1.1120	27.5	50.59	1.2361	40.5	76.29	1.3913
2.	3.59	1.0141	15.	27.19	1.1163	28.	51.55	1.2414	41.	77.32	1.3981
2.5	4.49	1.0177	15.5	28.10	1.1206	28.5	52.51	1.2468	41.5	78.35	1.4049
3.	5.39	1.0213	16.	29.03	1.1250	29.	53.47	1.2522	42.	79.39	1.4118
3.5	6.29	1.0249	16.5	29.95	1.1294	29.5	54.44	1.2576	42.5	80.43	1.4187
4.	7.19	1.0286	17.	30.87	1.1339	30.	55.47	1.2632	43.	81.47	1.4267
4.5	8.09	1.0323	17.5	31.79	1.1383	30.5	56.37	1.2687	43.5	82.51	1.4328
5.	9.00	1.0360	18.	32.72	1.1429	31.	57.34	1.2743	44.	83.56	1.4400
5.5	9.90	1.0397	18.5	33.65	1.1474	31.5	58.32	1.2800	44.5	84.62	1.4472
6.	10.80	1.0435	19.	34.58	1.1520	32.	59.29	1.2857	45.	85.68	1.4545
6.5	11.70	1.0473	19.5	35.50	1.1566	32.5	60.27	1.2915	45.5	86.74	1.4619
7.	12.61	1.0511	20.	36.44	1.1613	33.	61.25	1.2973	46.	87.81	1.4694
7.5	13.51	1.0549	20.5	37.37	1.1660	33.5	62.23	1.3032	46.5	88.81	1.4769
8.	14.42	1.0588	21.	38.30	1.1707	34.	63.22	1.3091	47.	89.96	1.4845
8.5	15.32	1.0627	21.5	39.24	1.1755	34.5	64.21	1.3151	47.5	91.03	1.4922
9.	16.23	1.0667	22.	40.17	1.1803	35.	65.20	1.3211	48.	92.12	1.5000
9.5	17.14	1.0706	22.5	41.11	1.1852	35.5	66.19	1.3272	48.5	93.21	1.5079
10.	18.05	1.0746	23.	42.05	1.1901	36.	67.19	1.3333	49.	94.30	1.5158
10.5	18.96	1.0787	23.5	42.99	1.1950	36.5	68.19	1.3395	49.5	95.40	1.5238
11.	19.87	1.0827	24.	43.94	1.2000	37.	69.19	1.3458	50.	96.51	1.5319
11.5	20.78	1.0868	24.5	44.88	1.2050	37.5	70.20	1.3521	50.5	97.62	1.5401
12.	21.69	1.0909	25.	45.83	1.2101	38.	71.20	1.3585	51.	98.73	1.5484
12.5	22.60	1.0951	25.5	46.78	1.2152	38.5	72.22	1.3649	51.5	99.65	1.5568

CHAPTER VI.

Determination of Cane-Sugar—Optical Methods.

POLARIZED LIGHT.

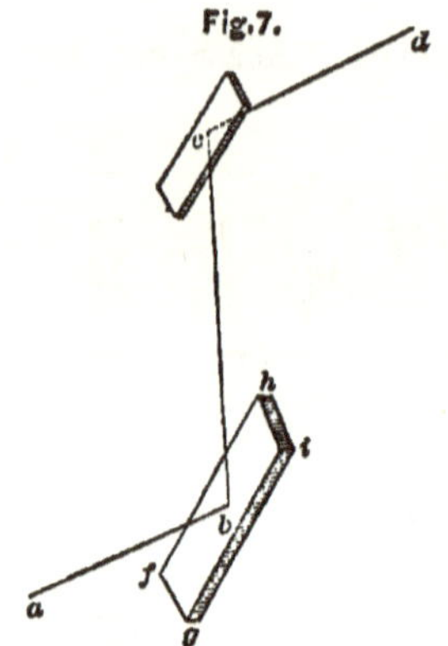

By Reflection.—When a ray of light, *a b*, Fig. 7, falls on a polished surface of glass (wood, ivory, leather, or other non-metallic substance), *f g h i*, inclined to it at an angle of 35° 25′, it is reflected, and the reflected ray acquires peculiar properties whereby it is said to be *polarized.* The change which has taken place in the light may be shown as follows: Let the polarized ray be received at *c* on a second reflecting surface, at the same angle as before. If the surfaces are parallel the ray is reflected; but if the second surface is caused to turn around *c b*, the intensity of the ray constantly diminishes, and when the reflecting planes are perpendicular to each other no light is reflected. If the rotation of the upper mirror be now continued the intensity of the ray gradually increases, and attains a maximum when the surfaces are again parallel. If the incident ray strikes at any other angle than that given the light is more or less polarized; but the greatest effect for glass is always obtained under the condition mentioned. The angle which the incident ray makes with the normal corresponds to the greatest effect for any substance, and is called the *polarizing angle.* For water it is 53° 11′; glass, 54° 35′; air, 45°; and quartz, 57° 32′.

By Refraction.—The phenomena of polarization are exhibited not only by reflection, but also by *refraction*, double or single. All doubly-refracting crystals have the property of polarizing light, and calc-spar may be selected as well illustrating this fact. When a ray of ordinary light passes through a crystal of calc-spar in any direction except that of the shorter diagonal of the rhomb, which is its optical axis, it is divided into two beams of equal intensity, the *ordinary* and the *extraordinary rays.* When the ordinary ray passes through a second rhomb of spar it again experiences double refraction, giving rise to two beams of unequal intensities. If the second crystal be rotated until the principal planes of the two coincide—that is, when they are in opposite or similar positions—the ordinary ray acquires its greatest intensity and the extraordinary ray disappears; continuing the rotation, the extraordinary ray reappears and increases in brightness, while the ordinary beam diminishes until the principal planes are perpendicular. When, however, the extraordinary ray suffers a second refraction by means of calc-spar, the converse to the above is exhibited. The two rays resulting from the double refraction are found to be polarized.

Among other crystalline bodies capable of polarizing light by double refraction may be mentioned tourmaline and selenite (crystallized sulphate of lime). Glass also, submitted to strains or pressure, becomes doubly-refracting. The plane in which a ray of polarized light, incident at the polarizing angle, is reflected or transmitted in the greatest degree, is called the *plane of polarization* of the ray.

When the polarization is produced by reflection the plane of polarization is identical with the plane of reflection.

The Nichol Prism.—A valuable device for producing polarized light, or analyzing it, is the *Nichol prism*, which consists of a rhomb of calc-spar slit along the plane passing through the shorter diagonal, and having the two halves cemented together again by Canada balsam, whose refractive index is intermediate between the ordinary and extraordinary indices of the crystal. Hence, when a ray of light, S C, Fig. 8, enters the prism, the ordinary ray experiences total reflection on the surface of the balsam, *a b*, and takes the direction C *d* O, and is refracted out of the crystal; while the extraordinary ray, C *e*, emerges alone. The Nichol prism has the advantages of perfect transparency and a very complete polarizing effect.

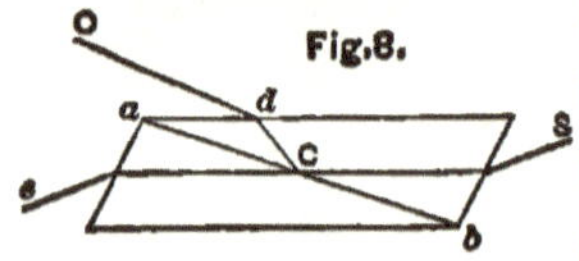

Elliptical, Circular, and Plane Polarization.—In accordance with the principles of the undulatory theory, when the ether particles that make up a beam of polarized light vibrate in parallel straight lines, the ray is said to be *plane polarized;* when the particles describe ellipses around their positions of rest, the planes of the ellipses being perpendicular to the ray and the axes parallel, the light is *elliptically polarized.* A particular case of the latter is when the axes of the ellipses become parallel, when *circular polarization* is produced. When a ray of light in this condition is refracted by a Nichol prism and viewed through an analyzer, the rotation of the latter causes no change in the intensity. Circularly-polarized light is not, however, identical with ordinary light, as may be proved by the interposition of a plate of selenite between the polarizer and analyzer, when the light becomes elliptically polarized.

Rotation of the Plane of Polarization.—Crystals of quartz, calc-spar, and tartaric acid can cause a rotation of the polarization plane around its axis. If a plate of quartz, cut perpendicular to its axis, is placed between the analyzer and polarizer, color is exhibited, the tints changing in the order of the colors of the spectrum as the analyzer is turned. With monochromatic light it is found that when the prisms are adjusted to produce total extinction of light, and the quartz introduced in the path of the ray, the light is partially restored, but that on rotating the analyzer again total extinction is produced. The angle through which it is necessary to turn the analyzer to produce this effect represents the angular rotation which the plane of polarization has experienced. There are two varieties of quartz, known as right and left handed—the one rotating the plane of polarization to the right and the other to the left. Fig. 9 represents the rotation of the plane of

Fig. 9.

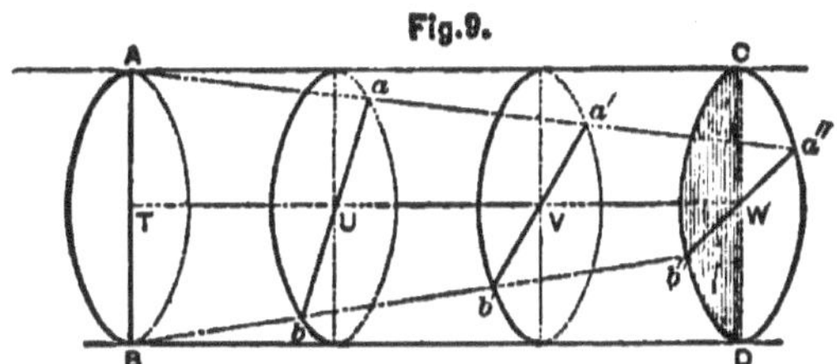

polarization: the plane A B, originally perpendicular, suffers successive rotations to $a\ b$, $a'\ b'$, and $a''\ b''$, the angle C W a'' being the final angle of rotation.

Malus has established the following laws in regard to rotatory polarization:

I. *The amount of rotation is proportional to the thick-*

ness of the quartz. II. *The rotation of the plane of polarization varies for the different rays of the spectrum, increasing with the refrangibility of the light.* With a plate of quartz one millimetre thick the rotations obtained for different colors were:

Red.............	19°	Blue............	32°
Orange...........	21°	Indigo..........	36°
Yellow...........	23°	Violet..........	41°
Green............	28°		

Specific Rotatory Power.—When the polarizer and analyzer are so placed to each other that their principal sections are parallel, and a quartz plate 3.75 mm. thick is interposed in the path of the polarized ray, a peculiar tint is produced. It is a delicate rose-purple, but changes quickly into red or violet by the slightest movement in the position of the analyzer, the alteration of color being much more rapid and decided than for any other shade or color. It is called the *transition tint* (*teinte de passage*), and in measurements of the rotative power of various bodies this is often taken as a standard. The rotatory power of liquids is directly as the length of the column through which the ray passes, and also as the quantity of active substance dissolved, if it is a solution. If e be the amount of substance dissolved in a unit of weight of the solution, l the length of the liquid column, and a the observed angle of rotation for any particular color, as the transition tint, the angle of rotation for the unit of length will be $\frac{a}{e\,l}$; but, as the solution of the optically active body is often attended with alteration of volume, it is desirable, in order to obtain an expression independent of such irregularities, to refer the observed angle of deviation to a hypothetical unit of density—that is, to divide the quantity $\frac{a}{e\,l}$ by the density, g, of the solution. The expression $[a]\,j = \frac{a}{e\,l\,g}$ is called the *specific rotatory power*, and represents the angle of deviation which the pure substance, in a column of the unit of length and density 1, would impart to the ray corresponding to the transition tint. For instance, a solution containing .155 gramme of cane-sugar to 1 gramme of liquid has a specific gravity of 1.06, and deflects the polarized ray for the transition tint 24° in a tube 20 mm. long. The specific rotatory power is, therefore,

$$[a]\,j = \frac{24}{.155 \times 20 \times 1.06} = 7.3°.$$

$[a]$ is the expression for the specific rotatory power in general; a letter affixed shows the particular ray of the spectrum at which the deviation was observed: thus, $[a]$ D and $[a]$ j are the expressions for the line D of the spectrum, and for the mean yellow ray, or transition tint, respectively. The minus-sign is prefixed to the degree when the substance rotates to the left.

The Polariscope.—The apparatus for determining the rotatory power is called a polariscope, and consists of an arrangement carrying two Nichol prisms properly placed to serve as analyzer and polarizer, having a space between them, so that a tube, provided with glass plates at its ends and filled with the solution to be examined, may be interposed in the path of the polarized ray. In front of the polarizer is inserted a quartz plate 3.75 mm. thick, so that when the prisms are adjusted with their principal planes parallel the transition tint is visible. The interposition of the active substance in the tube causes the color to change, and the amount of rotation of the analyzer necessary to restore the transition tint measures the angle of rotation of the body under examination, from which, with the data given, the specific rotatory power may be calculated. The instruments to be described furnish more elaborate and accurate means of determining the specific rotatory power.

Many organic bodies have the power of deviating the plane of polarization. Among them may be mentioned, DEVIATING TO THE RIGHT, cane-sugar, dextrose, milk-sugar, dextrin, camphor, asparagine, cinchonine, quinidine, narcotine, tartaric, camphoric, and aspartic acids, oil of lemons, and castor-oil; TO THE LEFT, levulose, starch, albumen, amygdalin, quinine, nicotine, strychnine, brucine, morphine, codeine, malic acid, oil of turpentine, and oil of valerian.

Optical Saccharimeters.—The property that solutions of cane-sugar have of deviating the polarized ray in a fixed and definite degree has been made the basis of various instruments constructed for the purpose of quantitatively estimating that body. These instruments are called *optical saccharimeters*, *polariscopes*, or *polarimeters*. Those treated of in this work are as follows: Mitscherlich's, the Soleil-Duboscq, the Soleil-Ventzke, Wild's Polaristrobometer, together with Duboscq's, Laurents's, and Schmidt and Haensch's modifications of the *saccharimètre à pénombre* of Jellett.

MITSCHERLICH'S SACCHARIMETER.

This instrument consists of two Nichol prisms, enclosed in brass tubes supported on a cast-iron foot by means of a bar, by which the upper part may be made to slide to and fro (Fig. 10). The tube *b* contains the polarizer, and it may be made to turn on its axis, being kept in any desired position by a screw at *i*. The tube *a*, containing the analyzer, is also capable of rotating, and has an arm attached, as well as a pointer which measures the amount of rotation upon a fixed graduated circle of brass. The graduation of the circle is in degrees from 0° to 360°. There is a space be-

Fig. 10.

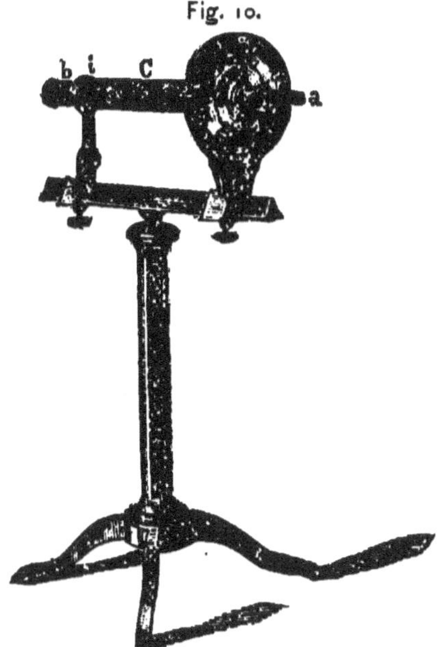

tween *a* and *b* for the reception of the tube C, which is exactly 200 mm. long and designed to hold the saccharine solution. This observation-tube is made of brass, and closed at each end by a screw-cap having a small orifice in its centre; glass plates are placed between the cap and the ground ends of the tube, so as to make a tight joint and to allow the light to pass through the axis of the tube.

The theory of the apparatus is very simple: the light entering by the first prism being polarized, on passing through the sugar solution has its plane deviated to the right; the prisms having their principal sections parallel, it becomes necessary to turn the analyzer through a certain angle corresponding to the strength of the solution, in order to compensate for the rotatory effect of the sugar.

To adjust the instrument for use it is important to fix correctly the zero-point, and that on the scale corresponding to 100 per cent. of cane-sugar. This is done as follows:

For the Zero—The pointer is turned to 0° on the scale, a gas or oil lamp being placed behind the apparatus in such a position that the light may shine through its axis, and the observation-tube, filled with water, having been put in place, *i* is unscrewed so as to allow the tube *b* to turn freely, the eye being placed at *a*. If the apparatus is not set correctly at the time of observation, a colored field will be seen, and the tube *b* must be turned until the field gradually darkens and finally presents the appearance of a round disk with an intense vertical black band in the centre, gradually shading equally on both sides to a lighter tint, and appearing dark green or yellowish at the extreme distance from the centre. When the field presents the above appearance the rotation of the tube *b* is suspended, and *i* is screwed down so as to secure it. Now, with the apparatus thus set, if *a* be turned by means of the index, the field becomes gradually lighter until the pointer indicates 90°, when it is at its maximum brightness; if the turning be continued the field darkens again, and at 180° it presents the same appearance as at 0°; this may be used as a control experiment for the correct adjustment of the zero-point.

If, when the instrument is properly adjusted, and the pointer stands at 0° on the scale, a colorless solution of cane-sugar be placed in the observation-tube, the field of the saccharimeter loses its dark color and shows a yellowish tint, owing to the fact that the plane of polarization has been altered by the sugar solution; on turning the analyzer in *a*, the field passes through a series of chromatic changes in the following order: yellow, green, blue, violet, red, orange. *To adjust the point corresponding to* 100 *per cent.* of sugar, a solution of 15 grammes pure, dry cane-sugar is made by dissolving in water and diluting to 100 c.c.; this is placed in the tube and the analyzer turned. The field passes through a series of colors as above until the normal spectrum of the apparatus is obtained, which presents an appearance as follows—viz.: the right half of the colored circle must appear of a pure blue; the centre has a line of violet, which shades off imperceptibly into red on the left. If the instrument correctly indicates at the point for 100 per cent. of sugar, the above appearance of the field is seen when the index of the scale is at 20°.

Use of the Instrument.—For use in testing saccharine products 15 grms. is taken, dissolved in water, and diluted to 100 c.c. After decolorization with lead solution, and filtering, some of the clear solution is placed in the observation-tube, and the analyzer turned by means of the arm attached, until the normal spectrum is obtained. The reading of the scale, multiplied by five, gives the percentage of sugar. It is evident that, when the degree of coloration of the material to be tested will admit, any multiple of the normal quantity may be taken and the solution made up to 100 c.c. The factor for multiplying the reading will be correspondingly less. With weak sugar solutions as much as 75 grms. may be weighed, in which case the reading of the instrument gives directly the percentage.

Value as a Saccharimeter.—The chief, and indeed almost fatal, objection to the Mitscherlich apparatus as an instrument of precision is that, in the majority of cases, the actual readings of the scale have to be multiplied by a large factor. Owing to the introduction of more accurate polarizing apparatus, the Mitscherlich instrument is now comparatively little used.

THE SOLEIL-DUBOSCQ SACCHARIMETER.

Biot, early in this century, investigated the principles of circular polarization, and especially the power which quartz plates have of rotating the plane of polarized light. He constructed the polariscope for measuring the rotatory quality of various substances, which, with the aid of calculation, was capable of quantitatively estimating sugar.

Clerget, following up the researches of Biot, devised a method of determining cane-sugar which is essentially that now employed with the Soleil saccharimeter. The method is Clerget's, the instrument is Soleil's.* The apparatus has been improved by Duboscq,† the successor of Soleil, and in its present form is called the saccharimeter of Soleil-Duboscq.

The Instrument.—The following is mainly Terreil's excellent description: Figure 11 represents the apparatus, which consists of two metal tubes mounted on an appropriate stand. The light enters at H by a circular opening of about 3 mm. diameter, and traverses the achromatic polarizing prism P; R is a plate of quartz, called the plate of double rotation, and is composed of two halves of equal thickness, *c d*, cut perpendicularly to the axis of

* Soleil, *Compt. Rend*, xxiv. 973. † Soleil et Duboscq, *ibid.*, xxxi. 248.

crystallization and joined together so that the line of separation is vertical. The half-disks have contrary rotations, the one being left-handed and the other right-handed. The light passing through T encounters Q, a quartz plate, either right or left handed, and of an arbitrary thickness. From Q the ray reaches K K′, which are two wedge-shaped

Fig. 11.

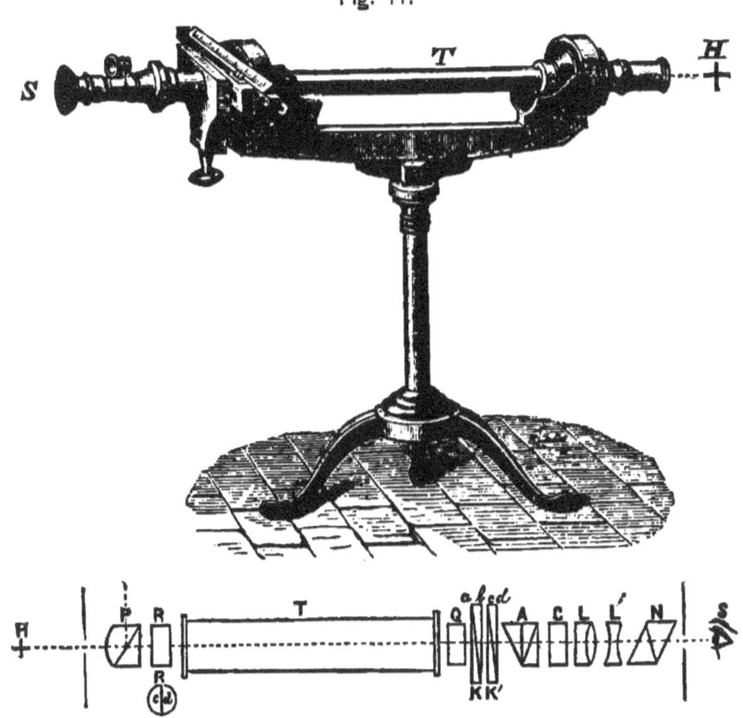

quartz plates, having the same kind of rotation, but different from that of Q. These plates are each fixed in a brass slide and covered with plane glass plates on each side to protect them from exterior injury or displacement.

By means of a rack-work and pinion, to which is fixed the milled head, the slides may be made to move to and

* The author is indebted to Dr. H. A. Mott for the above engraving.

fro in opposite directions while remaining parallel. By this arrangement, at will the thickness of the quartz through which the polarized ray has to pass may be varied. Finally the light passes to the analyzer A and the quartz plate C. The small Galilean telescope L L′ serves to render distinct the field of the instrument. The doubly-refracting prism A is so placed relatively to the diaphragm of the telescope that the passage of one of the rays transmitted by the polarizer is intercepted, so that but one passes, either the ordinary or the extraordinary ray, according as the plate R is 3.75 mm. or 7.5 mm. in thickness.

It is evident from the construction of the apparatus that on placing the eye at the ocular, S, there is seen the appearance of a luminous disk with a vertical line in the middle, produced by the junction of the quartz plates R. The sum of the thicknesses of the two prismatic quartz plates at a certain relative position is exactly equal to that of Q; and hence, as the rotations are in different senses, the one being left and the other right handed, or the reverse, it follows that they neutralize each other and produce no effect on the polarized ray. On looking into the instrument when thus adjusted it will be seen that the two half-disks of the field are of the same color. If now we interpose in the space T a tube containing a liquid having a rotatory power, immediately the uniformity of color between the two semi-disks is destroyed; this is due to the rotatory effect of the liquid, which destroys the mutual compensatory effect of R and the quartz wedges. For example, if the solution under examination consisted of cane-sugar, the deviation would be to the right, and this, with that of the right-handed plate of R, produces an inequality attended with the production of unequal color in the field. The field may be restored to uniformity by turning the screw, thereby increasing or decreasing the thickness of the quartz at K and compensating for the deviating effect of the liquid. This action of the *compensator* shows not only whether the solution of the substance examined is right or left rotating, but also the degree as measured by the thickness of quartz necessary to neutralize the deviation of the body examined. The latter is measured by means of a graduated scale fixed to one of the slides R R′ (Fig. 12), while upon the other is a mark serving as an indicator. The scale is graduated into degrees indicating percentages of sugar, on each side of the zero. A displacement of the scale equal to one division is equivalent to a rotative effect equivalent to that of a plate of quartz $\frac{1}{100}$ millimetre thick.

Fig. 12.

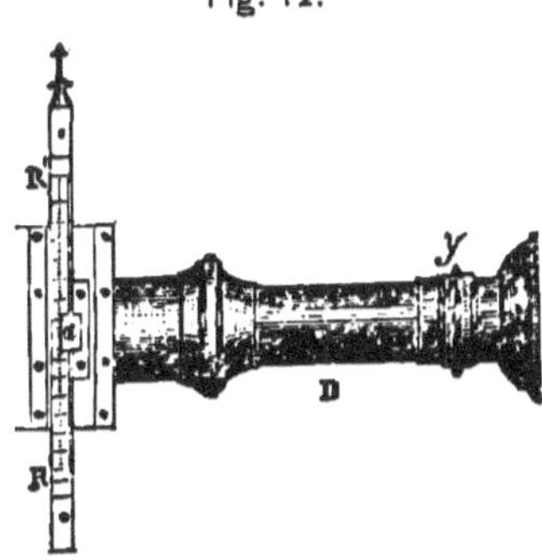

Soleil greatly improved his saccharimeter by placing in front of the ocular of the telescope a Nichol prism, N (Fig. 11), fixed in a movable case, which may be turned at will through an angle of 180°. This arrangement is called *the producer of sensitive tints*. The prism N destroys to a great extent the influence of the coloration in the liquids submitted to examination, and that of the light employed. It also permits us to obtain, by adjusting it to a certain position, the sensitive tint.

The tubes designed to contain the liquids to be tested consist entirely of brass, or glass enclosed in one of brass. The extremities of the tubes are ground, so as to be perfectly parallel with each other and to form a liquid-tight joint with the glass plates that cover them. Around the ends of the tubes there is a thread cut, by which brass caps, perforated in the centre, may be screwed on, a round plate of glass having been previously placed upon the end. The light can thus pass through the axis of the tube while it is filled with solution. An exterior view and section of these tubes may be seen in Fig. 13. The length of the

Fig. 13.

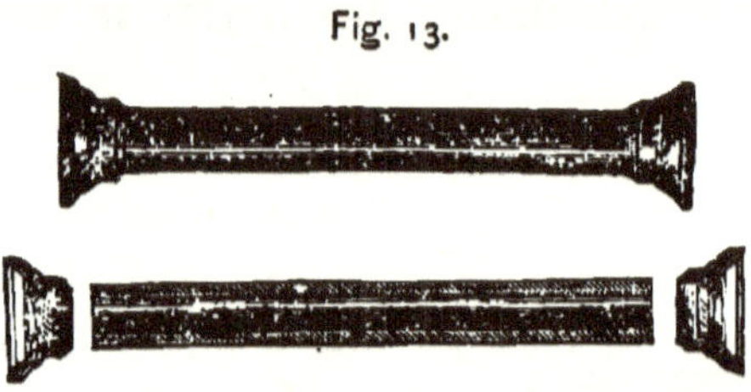

tubes is exactly 200 millimetres. The small movable tube containing the ocular to which the eye is placed, can be moved so as to adjust the focus in order to get the clearest view of the field. The collar on the ocular-tube, *y* (Fig. 12), which is connected with N, enables the operator to obtain the sensitive tint by rotating the prism.

Determination of the Zero-Point.—For this purpose the instrument is so placed that the light traverses its axis, and the observation-tube containing distilled water is put in position, as T in Fig. 11. The telescope is then focussed until a distinct view of the field is obtained. If the halves of the disk are different in color the milled head is turned either to the right or left, as may be necessary, until the colors appear to be perfectly identical on either side of the vertical line when the observation is taken; now the collar near the ocular is turned, and it will be perceived that the color of the field changes through red, blue, yellow, etc., until the sensitive tint is obtained, at which the previously appearing uniformity of the field, may be seen not to exist. A perfect uniformity may be made by turning the milled head cautiously again. The color of the sensitive tint varies somewhat with different observers, but for most persons it is the rose-violet, or where the lightest color of the spectrum (almost white) just begins to verge upon the red. By practising these manipulations the operator soon becomes skilled in the proper adjustment of the saccharimeter. When the field presents the appearance described, the zero of the scale ought to coincide with the indicator. Should this not be the case the two zeros may be made to agree by turning the screw-button (Fig. 12), placed near the end of the scale.

Manner of Using the Instrument.—To use the saccharimeter for the estimation of cane-sugar, a normal weight of 16.19 grms. is taken, dissolved in water, and the solution diluted up to 100 c.c., being suitably decolorized. When the observation-tube is filled with a solution thus prepared, and is placed in the instrument previously adjusted so that the field appears of a uniform tint, it will be seen that the uniformity is destroyed, and that the half-disks have different colors, one being complementary to the other. If now the milled head be turned until the equality of color is restored for the sensitive tint, the number of the scale to which the indicator points shows directly the percentage by weight of cane-sugar contained in the material examined.

A new instrument should be tested to see whether it

makes correct indications at the division of the scale reading 100 per cent., and whether the scale is correctly graduated, and the optical portions are in proper condition and adjustment. 16.19 grms. of pure, dry cane-sugar are taken, dissolved in water, and the solution made up to 100 c.c. This constitutes the *normal solution* for the saccharimeter, and should show 100° on the scale, the zero-point having been adjusted as previously described. A magnifying-glass accompanies the apparatus to assist in reading the scale.

Clerget's Method of Inversion.

The readings of the Soleil-Duboscq saccharimeter show directly the percentage of cane-sugar when no other optically active body is present. Such bodies are, however, often found in saccharine products submitted to the polariscopic test, particularly in beet syrups and juice. Under some conditions invert-sugar may also have a similar action, though this sugar is thought to be without action on the polarized ray when occurring in commercial saccharine products (see page 173).

As all of these substances have a specific rotatory power different from that of cane-sugar, deviating the plane either to the right or left, it follows that the reading of the saccharimeter for solutions containing such bodies must be incorrect as indicating cane-sugar, and the error will be in proportion to the amount of optically-active substance present.

Execution of the Process.—Clerget has devised a process for correcting the results of the saccharimeter when taken on solutions containing optically-active invert-sugar besides cane-sugar.* The direct titre is taken in the ordinary way, and a part of the solution remaining from this estimation is filled into a 50 c.c. flask (which is graduated to 50-55 c.c.) up to the 50 c.c. mark; then concentrated hydrochloric acid is added to 55 c.c., and the whole heated on a water-bath to 68°–75° for 10 to 15 minutes. This is sufficient to produce complete inversion of the cane-sugar present, while the invert-sugar is unacted on. After the liquid in the flask has attained the temperature of the surrounding air it is placed in the observation-tube and the reading taken. The sugar solution, while being heated with hydrochloric acid, is apt to become colored. The color can be readily removed by shaking the cold liquid with a very little bone-black. The observation-tube is of peculiar construction. It is larger than the ordinary, lined with glass, and has a tubule in the middle for the introduction of a thermometer-bulb in order to take the

Fig 14.

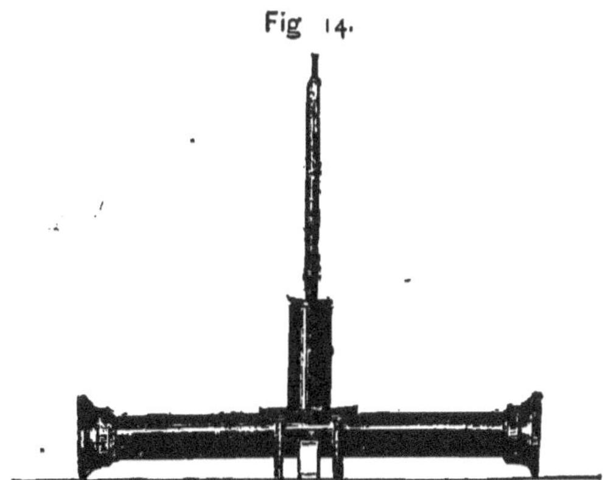

temperature of the liquid at the time of reading. Fig. 14 shows the arrangement.

* It must be remembered that the process is entirely inapplicable when any optically-active body is present besides cane or invert sugar, and also if the invert-sugar itself exists in an inactive condition as regards polarized light.

The tube is 220 mm. long, the increased length being to allow for the influence upon the saccharimetric reading made by the dilution of 10 per cent. on the addition of acid.

Calculation.—Clerget found that a solution of 16.35 grms. pure sugar in 100 c.c. of volume, which read $+100°$ in the saccharimeter, showed after inversion a rotation of 44° to the left at zero C.—a difference in the rotation of 144, due to the inversion. The optical rotation is much affected by the temperature of the solution after inversion, to the extent that the deviation diminishes by one-half of a degree (very nearly) of Soleil's scale for each degree Centigrade that the temperature is raised. At 0° C. the action is expressed by

$$T° = 144 - \tfrac{1}{2} T.$$

If S represents the sum or difference of the polariscopic readings before and after inversion, T the temperature of the inverted solution when polarized, and R the percentage of cane-sugar sought, then

$$144 - \tfrac{1}{2} T, : 100 :: S : R$$

$$288 - T : 200 :: S : R; \text{ whence}$$

$$R = \frac{200\ S}{288 - T}.$$

This formula, with the experimental data, will enable the operator to calculate the corrected percentage of cane-sugar.

Clerget's Table.—To save the trouble of this calculation Clerget has given a table, which will be found on pages 141, 142.* *Manner of Using the Table.*—When a liquid

* See also Tuchschmid, *Zeits. f. Rubenz. Ind.*, 1870, 649.

is tested in the saccharimeter, the degree of the scale has to be multiplied by 1.619 to give the number of grammes in a litre. This calculation the columns A and B enable us to dispense with. By finding in the column A the number of the scale read, the one corresponding under B shows the quantity sought. When the substance is submitted to inversion, the sum or difference* of the direct and indirect readings is taken, and the number nearest it in the column corresponding to the temperature at which the indirect reading was observed is sought. The horizontal line in which this number occurs is followed to the right, the quantity under A in this line being the corrected percentage of cane-sugar. For example:

I. Direct reading, + 38.7
Indirect " — 25 at 15° C.

Sum.......... 63.7

The nearest figure to the sum under 15° is 64.1, which corresponds to 47 per cent. of sugar.

II. Direct reading, + 90
Indirect " + 10 at 30° C.

Difference...... 80 = 62 per cent. sugar.

When the sum or difference does not correspond exactly to a number of the table in the temperature column, the sugar percentage should be taken for that next below and

* When the two readings are in the same sense—that is, both plus or both minus—the difference is taken; the sum is taken when they are in different senses.

above, and the average of two taken—as 63.7 under 15° C. is nearest to

62.8, corresponding to 46 per cent., and
64.1, " " 47 per cent.

Average............. 46.5

In all cases the results are calculated more exactly with the formula than by the table.

CLERGET'S TABLE FOR THE ANALYSES OF SACCHARINE SUBSTANCES.

SUMS AND DIFFERENCES OF THE DEVIATIONS, DIRECT AND INVERSE, TAKEN AT THE TEMPERATURES (CENT.)																										Per ct. sought.	
10°	11°	12°	13°	14°	15°	16°	17°	18°	19°	20°	21°	22°	23°	24°	25°	26°	27°	28°	29°	30°	31°	32°	33°	34°	35°	By wt. A	By vol. B
																											Gram.
1.4	1.4	1.4	1.4	1.4	1.4	1.4	1.3	1.3	1.3	1.3	1.3	1.3	1.3	1.3	1.3	1.3	1.3	1.3	1.3	1.3	1.3	1.3	1.3	1.3	1.3	1	1.64
2.8	2.8	2.8	2.7	2.7	2.7	2.7	2.7	2.7	2.7	2.7	2.7	2.6	2.6	2.6	2.6	2.6	2.6	2.6	2.6	2.6	2.6	2.6	2.6	2.6	2.6	2	3.29
4.2	4.1	4.1	4.1	4.1	4.1	4.1	4.1	4.0	4.0	4.0	4.0	4.0	4.0	4.0	3.9	3.9	3.9	3.9	3.9	3.9	3.8	3.8	3.8	3.8	3.8	3	4.94
5.6	5.5	5.5	5.5	5.5	5.5	5.4	5.4	5.4	5.4	5.4	5.3	5.3	5.3	5.3	5.3	5.2	5.2	5.2	5.2	5.2	5.1	5.1	5.1	5.1	5.1	4	6.58
6.9	6.9	6.9	6.9	6.8	6.8	6.8	6.8	6.7	6.7	6.7	6.7	6.6	6.6	6.6	6.6	6.5	6.5	6.5	6.5	6.4	6.4	6.4	6.4	6.3	6.3	5	8.23
8.3	8.3	8.3	8.2	8.2	8.2	8.2	8.1	8.1	8.1	8.0	8.0	8.0	7.9	7.9	7.9	7.9	7.8	7.8	7.8	7.7	7.7	7.7	7.7	7.6	7.6	6	9.88
9.7	9.7	9.7	9.6	9.6	9.5	9.5	9.5	9.4	9.4	9.4	9.3	9.3	9.3	9.2	9.2	9.2	9.1	9.1	9.1	9.0	9.0	9.0	8.9	8.9	8.8	7	11.52
11.1	11.1	11.0	11.0	11.0	10.9	10.9	10.8	10.8	10.8	10.7	10.7	10.6	10.6	10.6	10.5	10.5	10.4	10.4	10.4	10.3	10.3	10.2	10.2	10.2	10.1	8	13.17
12.5	12.5	12.4	12.4	12.3	12.3	12.2	12.2	12.1	12.1	12.1	12.0	12.0	11.9	11.9	11.8	11.8	11.7	11.7	11.6	11.6	11.6	11.5	11.5	11.4	11.4	9	14.82
13.9	13.8	13.8	13.7	13.7	13.6	13.6	13.5	13.5	13.4	13.4	13.3	13.3	13.2	13.2	13.1	13.1	13.0	13.0	12.9	12.9	12.8	12.8	12.7	12.7	12.6	10	16.47
15.3	15.2	15.2	15.1	15.1	15.0	14.0	14.9	14.8	14.8	14.7	14.7	14.6	14.6	14.5	14.5	14.4	14.3	14.3	14.2	14.2	14.1	14.1	14.0	14.0	13.9	11	18.11
16.7	16.6	16.6	16.5	16.4	16.4	15.3	16.3	16.2	16.1	16.1	16.0	16.0	15.9	15.8	15.8	15.7	15.7	15.6	15.5	15.5	15.4	15.4	15.3	15.2	15.2	12	19.76
18.1	18.0	17.9	17.9	17.8	17.7	16.7	17.6	17.5	17.5	17.4	17.3	17.3	17.2	17.2	17.1	17.0	17.0	16.9	16.8	16.8	16.7	16.6	15.6	16.5	16.4	13	21.41
19.5	19.4	19.3	19.2	19.2	19.1	19.0	19.0	18.9	18.8	18.8	18.7	18.6	18.5	18.5	18.4	18.3	18.3	18.2	18.1	18.1	18.0	17.9	17.8	17.8	17.7	14	23.05
20.8	20.8	20.7	20.6	20.5	20.5	20.4	20.3	20.2	20.2	20.1	20.0	19.9	19.9	19.8	19.7	19.6	19.6	19.5	19.4	19.3	19.3	19.2	19.1	19.0	19.0	15	24.70
22.2	22.2	22.1	22.0	21.9	21.8	21.8	21.7	21.6	21.5	21.4	21.4	21.3	21.2	21.1	21.0	21.0	20.9	20.8	20.7	20.6	20.6	20.5	20.4	20.3	20.2	16	26.35
23.6	23.5	23.5	23.4	23.3	23.2	23.1	23.0	22.9	22.9	22.8	22.7	22.6	22.5	22.4	22.3	23.3	22.2	22.1	22.0	21.9	21.8	21.7	21.7	21.6	21.5	17	28.00
25.0	24.9	24.8	24.7	24.7	24.6	24.5	24.4	24.3	24.2	24.1	24.0	23.9	23.8	23.8	23.7	23.6	23.5	23.4	23.3	23.2	23.1	23.0	22.9	22.9	22.8	18	29.64
26.4	26.3	26.2	26.1	26.0	25.9	25.8	25.7	25.6	25.5	25.5	25.4	25.3	25.2	25.1	25.0	24.9	24.8	24.7	24.6	24.5	24.4	24.3	24.2	24.1	24.0	19	31.29
27.8	27.7	27.6	27.5	27.4	27.3	27.2	27.1	27.0	26.9	26.8	26.7	26.6	26.5	26.4	26.3	26.2	26.1	26.0	25.9	25.8	25.7	25.6	25.5	25.4	25.3	20	32.94
29.2	29.1	29.0	28.9	28.8	28.7	28.6	28.4	28.3	28.2	28.1	28.0	27.9	27.8	27.7	27.6	27.5	27.4	27.3	27.2	27.1	27.0	26.9	26.8	26.7	26.6	21	34.58
30.6	30.5	30.4	30.2	30.1	30.0	29.9	29.8	29.7	29.6	29.5	29.4	29.3	29.1	29.0	28.9	28.8	28.7	28.6	28.5	28.4	28.3	28.2	28.0	27.9	27.8	22	36.23
32.0	31.8	31.7	31.6	31.5	31.4	31.3	31.2	31.0	30.9	30.8	30.7	30.6	30.5	30.4	30.2	30.1	30.0	29.9	29.8	29.7	29.5	29.5	29.3	29.2	29.1	23	37.88
33.4	33.2	33.1	33.0	32.9	32.8	32.6	32.5	32.4	32.3	32.2	32.0	31.9	31.8	31.7	31.6	31.4	31.3	31.2	31.1	30.9	30.8	30.7	30.6	30.5	30.4	24	39.53
34.7	34.6	34.5	34.4	34.2	34.1	34.0	33.9	33.7	33.6	33.5	33.4	33.2	33.1	33.0	32.9	32.7	32.6	32.5	32.4	32.2	32.1	32.0	31.9	31.7	31.6	25	41.17
36.1	36.0	35.9	35.7	35.6	35.5	35.4	35.2	35.1	35.0	34.8	34.7	34.6	34.4	34.3	34.2	34.1	33.9	33.8	33.7	33.5	33.4	33.3	33.1	33.0	32.9	26	42.82
37.5	37.4	37.3	37.1	37.0	36.8	36.7	36.6	36.4	36.3	36.2	36.1	35.9	35.8	35.6	35.5	35.4	35.2	35.1	35.0	34.8	34.7	34.6	34.4	34.3	34.1	27	44.47
38.9	38.8	38.6	38.5	38.4	38.2	38.1	37.9	37.8	37.7	37.5	37.4	37.2	37.1	37.0	36.8	36.7	36.5	36.4	36.3	36.1	36.0	35.8	35.7	35.6	35.4	28	46.11
40.3	40.2	40.0	39.9	39.7	39.6	39.4	39.3	39.1	39.0	38.9	38.7	38.6	38.4	38.3	38.1	38.0	37.8	37.7	37.5	37.4	37.3	37.1	37.0	36.8	36.7	29	47.75
41.7	41.5	41.4	41.2	41.1	40.9	40.8	40.6	40.5	40.3	40.2	40.0	39.9	39.7	39.6	39.4	39.3	39.1	39.0	38.8	38.7	38.5	38.4	38.2	38.1	37.9	30	49.41
43.1	42.9	42.8	42.6	42.5	42.3	42.2	42.0	41.8	41.7	41.5	41.4	41.2	41.1	40.9	40.8	40.6	40.4	40.3	40.1	39.9	39.8	39.7	39.5	39.4	39.2	31	51.06
44.5	44.3	44.2	44.0	43.8	43.7	43.5	43.4	43.2	43.0	42.9	42.7	42.6	42.4	42.2	42.1	41.9	41.8	41.6	41.4	41.3	41.1	41.0	40.8	40.6	40.5	32	52.70
45.9	45.7	45.5	45.4	45.2	45.0	44.9	44.7	44.5	44.4	44.2	44.0	43.9	43.7	43.6	43.4	43.2	43.1	42.9	42.7	42.6	42.4	42.2	42.1	41.9	41.7	33	54.35
47.3	47.1	46.9	46.7	46.6	46.4	46.2	46.1	45.9	45.7	45.6	45.4	45.2	45.0	44.9	44.7	44.5	44.4	44.2	44.0	43.9	43.7	43.5	43.3	43.2	43.0	34	56.00
48.6	48.5	48.3	48.1	47.9	47.8	47.6	47.4	47.2	47.1	46.9	46.7	46.5	46.4	46.2	46.0	45.8	45.7	45.5	45.3	45.1	45.0	44.8	44.6	44.4	44.3	35	57.64
50.0	49.9	49.7	49.5	49.3	49.1	49.0	48.8	48.6	48.4	48.2	48.1	47.9	47.7	47.5	47.3	47.2	47.0	46.8	46.6	46.4	46.3	46.1	45.9	45.7	45.5	36	59.29
51.4	51.2	51.1	50.9	50.7	50.5	50.3	50.1	49.9	49.8	49.6	49.4	49.2	49.0	48.8	48.6	48.5	48.3	48.1	47.9	47.7	47.5	47.4	47.2	47.0	46.8	37	60.94
52.8	52.6	52.4	52.2	52.1	51.9	51.7	51.5	51.3	51.1	50.9	50.7	50.5	50.3	50.2	50.0	49.8	49.6	49.4	49.2	49.0	48.8	48.7	48.4	48.3	48.1	38	62.58
54.2	54.0	53.8	53.6	53.4	53.2	53.0	52.8	52.6	52.4	52.3	52.1	51.9	51.7	51.5	51.3	51.1	50.9	50.7	50.5	50.3	50.1	49.9	49.7	49.5	49.3	39	64.23
55.6	55.4	55.2	55.0	54.8	54.6	54.4	54.0	54.0	53.8	53.6	53.4	53.2	53.0	52.8	52.6	52.4	52.2	52.0	51.8	51.6	51.4	51.2	51.0	50.8	50.6	40	65.88
57.0	56.8	56.6	56.4	56.2	56.0	55.8	55.5	55.3	55.1	54.9	54.7	54.5	54.3	54.1	53.9	53.7	53.5	53.3	53.1	52.9	52.7	52.5	52.3	52.1	51.9	41	67.53
58.4	58.2	58.0	57.7	57.5	57.3	57.1	56.9	56.7	56.5	56.3	56.1	55.9	55.6	55.4	55.2	55.0	54.8	54.6	54.4	54.2	54.0	53.8	53.5	53.3	53.1	42	69.17
59.8	59.5	59.3	59.1	58.9	58.7	58.5	58.3	58.0	57.8	57.6	57.4	57.2	57.0	56.8	56.5	56.3	56.1	55.9	55.7	55.5	55.2	55.0	54.8	54.6	54.4	43	70.82
61.2	60.9	60.7	60.5	60.3	60.1	59.8	59.6	59.4	59.2	59.0	58.7	58.5	58.3	58.1	57.9	57.6	57.4	57.2	57.0	56.8	56.5	56.3	56.1	55.9	55.7	44	72.47
62.5	62.3	62.1	61.9	61.6	61.4	[illegible]	61.0	60.7	60.5	60.3	60.1	59.8	59.6	59.4	59.2	58.9	58.7	58.5	58.3	58.0	57.8	57.6	57.4	57.1	56.9	45	74.11
63.9	63.7	63.5	63.2	63.0	62.8	62.6	62.3	62.1	61.9	61.6	61.4	61.2	60.9	60.7	60.5	60.3	60.0	59.8	59.6	59.3	59.1	58.9	58.6	58.4	58.2	46	75.76
65.3	65.1	64.9	64.6	64.4	64.1	63.9	63.7	63.4	63.2	63.0	62.7	62.5	62.3	62.0	61.8	61.6	61.3	61.1	60.9	60.6	60.4	60.2	59.9	59.7	59.4	47	77.41
66.7	66.5	66.2	66.0	65.8	65.5	65.3	65.0	64.8	64.6	64.3	64.1	63.8	63.6	63.4	63.1	62.9	62.6	62.4	62.2	61.9	61.7	61.4	61.2	61.0	60.7	48	79.06
68.1	67.9	67.6	67.4	67.1	66.9	66.6	66.4	66.1	65.9	65.7	65.4	65.2	64.9	64.7	64.4	64.2	63.9	63.7	63.4	63.2	63.0	62.7	62.5	62.2	62.0	49	80.70

SUMS AND DIFFERENCES OF THE DEVIATIONS, DIRECT AND INVERSE, TAKEN AT THE TEMPERATURES (CENT.)																										Per ct. sought.	
10°	11°	12°	13°	14°	15°	16°	17°	18°	19°	20°	21°	22°	23°	24°	25°	26°	27°	28°	29°	30°	31°	32°	33°	34°	35°	By wt. A	By vol. B
																											Gram.
69.5	69.2	69.0	68.7	68.5	68.2	68.0	67.7	67.5	67.2	67.0	66.7	66.5	65.2	66.0	65.7	65.5	65.2	65.0	64.7	64.5	64.2	64.0	63.7	63.5	63.2	50	82.35
70.9	70.6	70.4	70.1	69.9	69.6	69.4	69.1	68.8	68.6	68.3	68.1	67.8	67.6	67.3	67.1	66.8	66.5	66.3	66.0	65.8	65.5	65.3	65.0	64.8	64.5	51	84.00
72.3	72.0	71.8	71.5	71.2	71.0	70.7	70.5	70.2	69.9	69.7	69.4	69.2	68.9	68.6	68.4	68.1	67.9	67.6	67.3	67.1	66.8	66.6	66.3	66.0	65.8	52	85.64
73.7	73.4	73.1	72.9	72.6	72.3	72.1	71.8	71.5	71.3	71.0	70.7	70.5	70.2	70.0	69.7	69.4	69.2	68.9	68.6	68.4	68.1	67.8	67.6	67.3	67.0	53	87.29
75.1	74.8	74.5	74.2	74.0	73.7	73.4	73.2	72.9	72.6	72.4	72.1	71.8	71.5	71.3	71.0	70.7	70.5	70.2	69.9	69.7	69.4	69.1	68.8	68.6	68.3	54	88.94
76.4	76.2	75.9	75.6	75.3	75.1	74.8	74.5	74.2	74.0	73.7	73.4	73.1	72.9	72.6	72.3	72.0	71.8	71.5	71.2	70.9	70.7	70.4	70.1	69.8	69.6	55	90.59
77.8	77.6	77.3	77.0	76.7	76.4	76.2	75.9	75.6	75.3	75.0	74.8	74.5	74.2	73.9	73.6	73.4	73.1	72.8	72.5	72.2	72.0	71.7	71.4	71.1	70.8	56	92.23
79.2	79.0	78.7	78.4	78.1	77.8	77.5	77.2	76.9	76.7	76.4	76.1	75.8	75.5	75.2	74.9	74.7	74.4	74.1	73.8	73.5	73.2	73.0	72.7	72.4	72.1	57	93.88
80.6	80.3	80.0	79.7	79.5	79.2	78.9	78.6	78.3	78.0	77.7	77.4	77.1	76.8	76.6	76.3	76.0	75.7	75.4	75.1	74.8	74.5	74.2	73.9	73.7	73.4	58	95.53
82.0	81.7	81.4	81.1	80.8	80.5	80.2	79.9	79.6	79.3	79.1	78.8	78.5	78.2	77.9	77.6	77.3	77.0	76.7	76.4	76.1	75.8	75.5	75.2	74.9	74.6	59	97.17
83.4	83.1	82.8	82.5	82.2	81.9	81.6	81.3	81.0	80.7	80.4	80.1	79.8	79.5	79.2	78.9	78.6	78.3	78.0	77.7	77.4	77.1	76.8	76.5	76.2	75.9	60	98.82
84.8	84.5	84.2	83.9	83.6	83.3	83.0	82.6	82.3	82.0	81.7	81.4	81.1	80.8	80.5	80.2	79.9	79.6	79.3	79.0	78.7	78.4	78.1	77.8	77.5	77.2	61	100.47
86.2	85.9	85.6	85.2	84.9	84.6	84.3	84.0	83.7	83.4	83.1	82.8	82.5	82.1	81.8	81.5	81.2	80.9	80.6	80.3	80.0	79.7	79.4	79.0	78.7	78.4	62	102.12
87.6	87.2	86.9	86.5	86.3	86.0	85.7	85.4	85.0	84.7	84.4	84.1	83.8	83.5	83.2	82.8	82.5	82.2	81.9	81.6	81.3	80.9	80.6	80.3	80.0	79.7	63	103.76
89.0	88.6	88.3	88.0	87.7	87.4	87.0	86.7	86.4	86.1	85.8	85.4	85.1	84.8	84.5	84.2	83.8	83.5	83.2	82.9	82.6	82.2	81.9	81.6	81.3	81.0	64	105.41
90.3	90.0	89.7	89.4	89.0	88.7	88.4	88.1	87.7	87.4	87.1	86.8	86.4	86.1	85.8	85.5	85.1	84.8	84.5	84.2	83.8	83.5	83.2	82.9	82.5	82.2	65	107.06
91.7	91.4	91.1	90.7	90.4	90.1	89.8	89.4	89.1	88.8	88.4	88.1	87.8	87.4	87.1	86.8	86.5	86.1	85.8	85.5	85.1	84.8	84.5	84.1	83.8	83.5	66	108.70
93.1	92.8	92.5	92.1	91.8	91.4	91.1	90.8	90.4	90.1	89.8	89.4	89.1	88.8	88.4	88.1	87.8	87.4	87.1	86.8	86.4	86.1	85.8	85.4	85.1	84.8	67	110.35
94.5	94.2	93.8	93.5	93.2	92.8	92.5	92.1	91.8	91.5	91.1	90.8	90.4	90.1	89.8	89.4	89.1	88.7	88.4	88.1	87.7	87.4	87.0	86.7	86.4	86.0	68	112.00
95.9	95.6	95.2	94.9	94.5	94.1	93.8	93.5	93.1	92.8	92.5	92.1	91.8	91.4	91.1	90.7	90.4	90.0	89.7	89.3	89.0	88.7	88.3	88.0	87.6	87.3	69	113.64
97.3	96.9	96.6	96.2	95.9	95.5	95.2	94.8	94.5	94.1	93.8	93.4	93.1	92.7	92.4	92.0	91.7	91.3	91.0	90.6	90.3	89.9	89.6	89.2	88.9	88.5	70	115.29
98.7	98.3	98.0	97.6	97.3	96.9	96.6	96.2	95.8	95.5	95.1	94.8	94.4	94.1	93.7	93.4	93.0	92.6	92.3	91.9	91.6	91.2	90.9	90.5	90.2	89.8	71	116.94
100.1	99.7	99.4	99.0	98.6	98.3	97.9	97.6	97.2	96.8	96.5	96.1	95.8	95.4	95.0	94.7	94.3	94.0	93.6	93.2	92.9	92.5	92.2	91.8	91.4	91.1	72	118.59
101.5	101.1	100.7	100.4	100.0	99.6	99.3	98.9	98.5	98.2	97.8	97.4	97.1	96.7	96.4	96.0	95.6	95.3	94.9	94.5	94.2	93.8	93.4	93.1	92.7	92.3	73	120.23
102.9	102.5	102.1	101.7	101.4	101.0	100.6	100.3	99.9	99.5	99.2	98.8	98.4	98.0	97.7	97.3	96.9	96.6	96.2	95.8	95.5	95.1	94.7	94.3	94.0	96.6	74	121.88
104.2	103.9	103.5	103.1	102.7	102.4	102.0	101.6	101.2	100.9	100.5	100.1	99.7	99.4	99.0	98.6	98.2	97.9	97.5	97.1	96.7	96.4	96.0	95.6	95.2	94.9	75	123.53
105.6	105.3	104.9	104.5	104.1	103.7	103.4	103.0	102.6	102.2	101.8	101.5	101.1	100.7	100.3	99.9	99.6	99.2	98.8	98.4	98.0	97.7	97.3	96.9	96.5	96.1	76	125.17
107.0	106.6	106.3	105.9	105.5	105.1	104.7	104.3	103.9	103.6	103.2	102.8	102.4	102.0	101.6	101.2	100.9	100.5	100.1	99.7	99.3	98.9	98.6	98.2	97.8	97.4	77	126.82
108.4	108.0	107.6	107.2	106.9	106.5	106.1	105.7	105.3	104.9	104.5	104.1	103.7	103.3	103.0	102.6	102.2	101.8	101.4	101.0	100.6	100.2	99.8	99.4	99.1	98.7	78	128.47
109.8	109.4	109.0	108.6	108.2	107.8	107.4	107.0	106.6	106.2	105.9	105.5	105.1	104.7	104.3	103.9	103.5	103.1	102.7	102.3	101.9	101.5	101.1	100.7	100.3	99.9	79	130.12
111.2	110.8	110.4	110.0	109.6	109.2	108.8	108.4	108.0	107.6	107.2	106.8	106.4	106.0	105.6	105.2	104.8	104.4	104.0	103.6	103.2	102.8	102.4	102.0	101.6	101.2	80	131.76
112.6	112.2	111.8	111.4	111.0	110.6	110.2	109.7	109.3	108.9	108.5	108.1	107.7	107.3	106.9	106.5	106.1	105.7	105.3	104.9	104.6	104.1	103.7	103.2	102.9	102.5	81	133.41
114.0	113.6	113.2	112.7	112.3	111.9	111.5	111.1	110.7	110.3	109.9	109.5	109.1	108.6	108.2	107.8	107.4	107.0	106.6	106.2	105.9	105.4	105.0	104.5	104.1	103.7	82	135.06
115.4	114.9	114.5	114.1	113.7	113.3	112.9	112.5	112.0	111.6	111.2	110.8	110.4	110.0	109.6	109.1	108.7	108.3	107.9	107.5	107.2	106.5	106.2	105.8	105.4	105.0	83	136.70
116.8	116.3	115.9	115.5	115.1	114.7	114.2	113.8	113.4	113.0	112.6	112.1	111.7	111.3	110.9	110.5	110.0	109.6	109.2	108.8	108.5	107.9	107.5	107.1	106.7	106.3	84	138.35
118.1	117.7	117.3	116.9	116.4	116.0	115.6	115.2	114.7	114.3	113.9	113.5	113.0	112.6	112.2	111.8	111.3	110.9	110.5	110.1	109.7	109.2	108.8	108.4	107.9	107.5	85	140.00
119.5	119.1	118.7	118.2	117.8	117.4	117.0	116.5	116.1	115.7	115.2	114.8	114.4	113.9	113.5	113.1	112.7	112.2	111.8	111.4	111.0	110.5	110.1	109.6	109.2	108.8	86	141.65
120.9	120.5	120.1	119.6	119.2	118.7	118.3	117.9	117.4	117.0	116.6	116.1	115.7	115.3	114.8	114.4	114.0	113.5	113.1	112.7	112.3	111.8	111.4	110.9	110.5	110.0	87	143.29
122.3	121.9	121.4	121.0	120.6	120.1	119.7	119.2	118.8	118.4	117.9	117.5	117.0	116.6	116.2	115.7	115.3	114.8	114.4	114.0	113.6	113.1	112.6	112.2	111.8	111.3	88	144.94
123.7	123.2	122.8	122.4	121.9	121.5	121.0	120.6	120.1	119.7	119.3	118.8	118.4	117.9	117.5	117.0	116.6	116.1	115.7	115.2	114.9	114.4	113.9	113.5	113.0	112.6	89	146.59
125.1	124.6	124.2	123.7	123.3	122.8	122.4	121.9	121.5	121.0	120.6	120.1	119.7	119.2	118.8	118.3	117.9	117.4	117.0	116.5	116.2	115.6	115.2	114.7	114.3	113.9	90	148.23
126.5	126.0	125.0	125.1	124.7	124.2	123.8	123.3	122.8	122.4	121.9	121.5	121.0	120.6	120.1	119.7	119.2	118.7	118.3	117.8	117.5	115.9	116.5	116.0	115.6	115.1	91	149.88
127.9	127.4	127.0	126.5	126.0	125.6	125.1	124.7	124.2	123.7	123.3	122.8	122.4	121.9	121.4	121.0	120.5	120.1	119.6	119.1	118.8	118.2	117.8	117.3	116.8	116.4	92	151.53
129.3	128.8	128.3	127.9	127.4	126.9	126.5	126.0	125.5	125.1	124.6	124.1	123.7	123.2	122.8	122.3	121.8	121.4	120.9	120.4	120.1	119.5	119.0	118.6	117.1	117.6	93	153.18
130.7	130.2	129.7	129.2	128.8	128.3	127.8	127.4	126.9	126.4	126.0	125.5	125.0	124.5	124.1	123.6	123.1	122.7	122.2	121.7	121.4	120.8	120.3	119.8	119.4	118.9	94	154.82
132.0	131.6	131.1	130.6	130.1	129.7	129.2	128.7	128.2	127.8	127.3	126.8	126.3	125.9	125.4	124.9	124.4	124.0	123.5	123.0	122.6	122.1	121.6	121.1	120.6	120.2	95	156.47
133.4	133.0	132.5	132.0	131.5	131.0	130.6	130.1	129.6	129.1	128.6	128.2	127.7	127.2	126.7	126.2	125.8	125.3	124.8	124.3	123.9	123.4	122.9	122.4	121.9	121.4	96	158.12
134.8	134.3	133.9	133.4	132.9	132.4	131.9	131.4	130.9	130.5	130.0	129.5	129.0	128.5	128.0	127.5	127.1	126.6	126.1	125.6	125.2	124.6	124.2	123.7	123.2	122.7	97	159.76
136.2	135.7	135.2	134.7	134.3	133.8	133.3	132.8	132.3	131.8	131.3	130.8	130.3	129.8	129.4	128.9	128.4	127.9	127.4	126.9	126.5	125.9	125.4	124.9	124.5	124.0	98	161.41
137.6	137.1	136.6	136.1	135.6	135.1	134.6	134.1	133.6	133.1	132.7	132.2	131.7	131.2	130.7	130.2	129.7	129.2	128.7	128.2	127.8	127.2	126.7	126.2	125.7	125.2	99	163.06
139.0	138.5	138.0	137.5	137.0	136.5	136.0	135.5	135.0	134.5	134.0	133.5	133.0	132.5	132.0	131.5	131.0	130.5	130.0	129.5	129.0	128.5	128.0	127.5	127.0	126.5	100	164.71

The Method applied for Saccharimeters in General.—The underlying fact of Clerget's process—namely, that a sugar solution reading $+100°$ will, after the action of acids, show $-44°$, making a difference of 144° due to inversion—is general, and hence may be applied to the results of any saccharimeter. By the following method of proceeding, instead of the one described, fully as accurate results may be obtained with much less trouble, and only the observation-tubes used in ordinary work. The direct reading is taken, and from the normal solution remaining 50 c.c. are placed in a flask graduated to 50–55 c.c., acid being added to the upper mark, and the sugar inverted as previously described. After inversion the solution is allowed to cool, the evaporated water replaced, and the reading taken in the ordinary glass tube, the temperature from a thermometer placed near the saccharimeter being also observed. The reading is increased by ten per cent. Care must be taken to keep the temperature in the neighborhood of the instrument as uniform as possible, and to bring the solution to the same degree before filling into the tube. If these precautions are taken the temperature will not vary materially during the observation. The calculation is the same as that already given, either the formula or table being used.

THE SOLEIL-VENTZKE SACCHARIMETER.

This instrument differs in no essential from the one last described, though the mechanical construction has been greatly improved, the optical parts somewhat changed and arranged in a different manner. These improvements are due to Ventzke,* and later to Scheibler.† This saccharimeter, as now made by the best European makers,‡ is one of the most practically useful for the optical determination of cane-sugar, and is to be recommended in preference to the Soleil-Duboscq, though more expensive. Owing to its perfection in mechanical construction it is very easy to work with, and in regard to accuracy leaves nothing to be desired for all technical work.

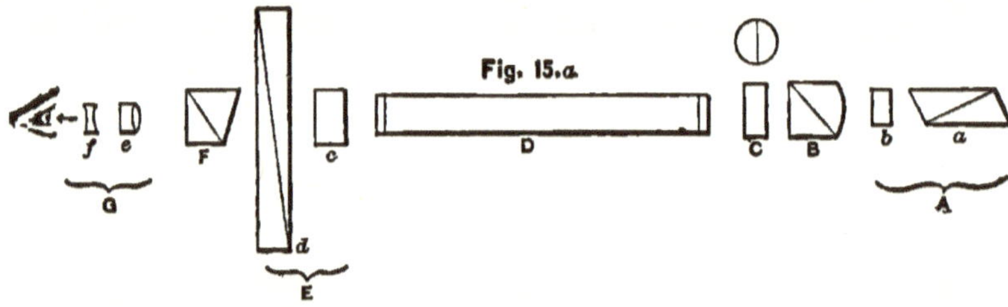

Description of the Instrument.—Fig. 15*a* shows the arrangement of the optical portions:

1. A is the regulator for changing the tints of the double quartz plates C. It consists of the Nichol *a*, and a quartz plate *b*, cut perpendicular to the axis of the crystal, both of which can be caused to rotate by appropriate means.

2. B, the polarizer, is an achromatic calc-spar prism. As its principal section is vertical, the extraordinary ray is totally reflected at the axis, and only the ordinary ray is transmitted. The convex surface turned towards A renders the rays parallel.

3. The double quartz plate C is precisely similar to that

---

* Ventzke, *Journ. f. Prk. Chemie*, xxv. 84, xxviii. 3.

† Scheibler, *Zeitschrift für Rubenz.*, 1870, 609.

‡ Dr. C. Scheibler, 24 Alexandrinen Str.; Schmidt u. Haensch, 4 Stallschreiber Str., Berlin.

of the Soleil-Duboscq apparatus. Its thickness may be either 3.75 or 7.50 mm.

4. The observation-tube, D.

5. The compensator, E, consists of the right-handed plate of quartz *c*, and the wedge-form plates *d*, which are of left-handed quartz, one of which is fixed and the other movable by means of a rack and pinion, to increase or diminish the thickness of crystal through which the polarized ray has to pass; *c* may be of left-handed quartz, but in that case the optical rotation of the wedges must be in an opposite sense.

Fig. 15*b*.

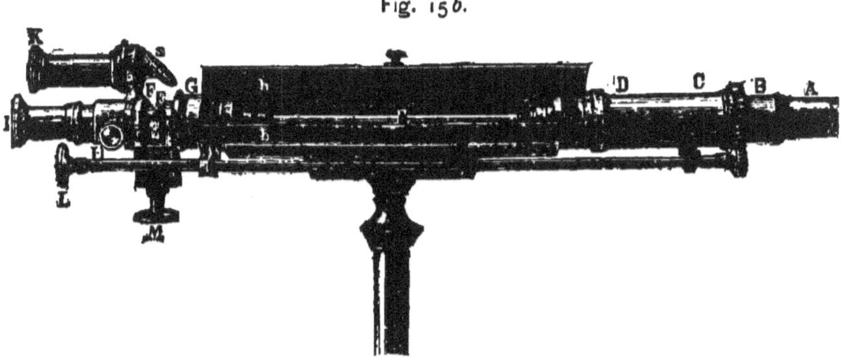

6. The analyzer, F, is an achromatic calc-spar prism, whose principal section must be parallel to that of the polarizer, B, when the thickness of the plate C is 3.75 mm., or perpendicular to it when the latter is double that thickness.

7. G is a small Galilean telescope, consisting of objective *e* and ocular *f*.

The general optical theory and manner of working with the Soleil-Ventzke is the same as that of the Soleil-Duboscq, and the reader is referred to the description of that instrument for these particulars. Only to points where they differ will particular attention be paid in this place.

Fig. 15*b* gives a complete perspective view of the instrument with the latest improvements introduced by Scheibler. A brass support standing on an iron tripod holds the main portion of the apparatus, the middle part of which consists of a japanned metal receptacle, *h*, for the observation-tube, provided with a hinged cover, which serves to shut out the light while an observation is taken. At one end of this is fixed a brass tube containing the double quartz plate, D, and the polarizer, C. To this tube is fastened a metal case, A B, arranged so as to be capable of turning freely upon its axis, and which, with the quartz plate *b* and the Nichol *a*, constitutes the *regulator* (Fig. 15 *a*). The regulator is rotated by a toothed wheel attached to it, actuating in a pinion fastened to a rod which terminates at the front of the instrument in a milled head, L, where it can be conveniently reached by the operator. At G is the compensator, and F E are the quartz wedges, each of which is secured in a strong brass frame and covered with parallel plates of glass on each side; F is fixed by two screws and carries the vernier, while E can move horizontally by means of a toothed rack on the lower portion of the brass frame, and a pinion moved by the milled head M; E carries the scale, which is graduated on both sides of zero. The scale and vernier are not shown in the plate; the latter reads to tenths of one per cent. In order to read the scale, a horizontally-placed telescope, K, is screwed on to the apparatus, and the light from the scale is reflected into it by the mirror, S. The analyzer may be turned by a key, so that it can be put into proper relation to the polarizer, if necessary. The key also may be used to adjust the zero of the scale to that of the vernier by means of a screw in F not shown in the plate.

Adjustment of Prisms.—If, by any cause, the analyzer and polarizer are not in perfect adjustment towards each other, which is shown by the fact that for any position of the plates F E there is no equality of tint on both sides of the vertical line in the field of the apparatus, the adjustment must be made. For this purpose E is removed by turning M until it can be taken out; then the screws that hold F are unscrewed, and this plate also removed; and finally the compensation-plate is displaced. Now, with the cover closed, an observation is taken, and, by means of the key, H is turned until the field gives the normal spectrum; the key is then taken out and the parts replaced as they were before. Finally, the zero of the vernier is adjusted to correspond to that of the scale by an observation taken with an empty tube, by turning the screw on F by means of the key. When the apparatus is thus adjusted it will give correct indications for all points of the scale, provided the latter is equally divided, and the instrument is not essentially faulty in construction.

On the scale of the original Soleil-Ventzke saccharimeter a solution of pure cane-sugar of a density 1.10 at 17½° C., observed in a tube 200 mm. long, reads 100°. It has been experimentally proved that such a solution contains 26.048 grms. cane-sugar in 100 c.c. Hence, if 26.048 grms. of sugar be weighed, dissolved in water, and the solution diluted to 100 c.c., the result, as read in the saccharimeter, would be the same as if the solution were prepared of the normal density. Ventzke used a special areometer (page 106), giving the densities required with great accuracy. The method of direct weighing the normal quantity is now used altogether in place of the earlier one with the areometer.

Method of Using the Apparatus.—The material to be tested is weighed in a tared dish provided with a counterpoise. Any balance will serve that weighs quickly and accurately to .010 grm., as an error of this quantity makes a difference in the reading of less than $\frac{1}{25}$ of one degree on the scale. **The observation-tubes** are of glass, respectively 200 and 100 mm. long, furnished with screw-caps and glass plates to close the ends (page 134). Glass tubes are objectionable not only on account of their fragility, but also because the brass screw-threads at the end frequently become loose, the effect of which is to lengthen unduly the column of liquid under observation, rendering the reading too high. A brass tube of the same form and dimensions may be used with great convenience. The only objection which can be urged against the latter is that the coefficient of the linear expansion of brass is greater than that of glass, and consequently variations of temperature in altering the length of the tube would give rise to error. This objection is, however, not well founded, as it can be proved by calculation that in the most extreme cases the maximum error for a tube 200 mm. long corresponds to less than .04 per cent. sugar.

The shorter tube (100 mm.) should only be used when it is impossible or inconvenient to get a solution light enough to read in the longer one. When the readings are taken with the former they are to be doubled to make them indicate percentages of sugar.

To Test the Correctness of the Saccharimeter.—When a new saccharimeter is obtained, or the operator uses one with whose antecedents he is not familiar, it should be thoroughly examined. First the observation-tubes should be measured with care to see whether they are of standard length. For this purpose a reliable metal or ivory rule should be procured, graduated into millimetres. The tube may be measured by a pair of accurate calipers, which should be perfectly adjusted to the ends of the tube, and then applied to the standard rule. If after several trials the tube is found to be too long, it must be ground down to the right length with oil and emery on a thick glass plate; if, on the other hand, it is too short, it must be rejected, or a correction made for the readings taken with it as follows: Suppose, for example, a tube measured 199 mm.; as the readings of the saccharimeter are directly proportional to the length of the column of saccharine liquid, and 200 mm. corresponds to 100°, we have

$$200 : 199 :: 100 : x = 99.5°.$$

The various adjustments for the zero and 100° point of the scale are made in an entirely similar manner to those for the Soleil-Duboscq saccharimeter; it is to be understood that before the adjustment at 0° is made it has been ascertained whether the analyzer and polarizer are in proper relation, and if they are not they must be corrected according to the directions already given. The 100° point of the scale is tested by dissolving 26.048 grammes of pure cane-sugar in water, diluting to 100 c.c., and taking a careful observation with the solution thus obtained in the 100 mm. and 200 mm. tubes; their readings should be exactly 50° and 100° respectively. The correctness of the division of the scale is best tested in the laboratory by weighing indefinite quantities of pure sugar, less than the normal weight, dissolving in water, diluting to 100 c.c., and polarizing—thus, if 20.50 grammes of sugar are taken, then

$$26.05 : 20.5 :: 100 : x$$

$$\frac{20.5 \times 100}{26.05} = 78.7,$$

which is the division of the scale that the solution should indicate. If the indications for various points are different from those which the amounts of sugar taken should give, while the 0° and 100° point is correct, the scale is not properly divided; if the error exists to any considerable extent, and at different parts of the scale, the instrument should be rejected, or a new scale obtained for it. Scheibler * has given a method for correcting the scale, which he calls the "*Hundert Polarisation*," and which consists in first obtaining the polarization of a raw sugar or other saccharine material, and then calculating the amount necessary to be weighed to polarize 100; as, for example, a sugar polarizing 85 would require $\frac{26.05 \times 100}{85} = 30.65$ grammes to be taken for the test to show a saccharimetric reading of 100. If a number of points on the scale, distributed from 0° to 100°, are found to be correct, the saccharimeter may be accepted as reliable. The troublesome operation of preparing pure sugar and making solutions of different strengths to test the correctness of the scale may be dispensed with by employing quartz plates of various thicknesses, and consequently whose rotatory powers correspond to sugar solution of different strengths. Such plates are made by Dr. Scheibler, of Berlin, for use on almost every part of the scale from 38° to 100°; it is only neces-

* *Zeits. f. Zuckerfabr. des deutsch. Reiches*, xxi. 320.

sary to place them in the end of the observation-tube and to proceed as if a sugar solution was to be examined.

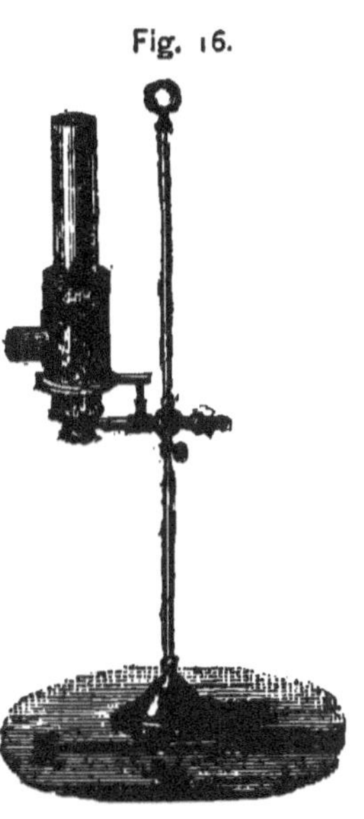

Fig. 16.

Source of Light.—The source of light for use with this and other saccharimeters not requiring the monochromatic flame may be either a good Argand oil-lamp such as shown in Fig. 16, or an ordinary Argand gas-burner.

A saccharimeter is best mounted for laboratory work in a wooden case of suitable dimensions, placed in the darkest part of the room and supported on brackets, or in any other way. The end of the polariscope should be placed at least six centimetres from the source of light to avoid the danger of softening the cement used in keeping the prisms of the apparatus in place. The case intended for the reception of the instrument may have a hinged top that can be thrown back when the apparatus is in use, and also a door in front provided with a lock and key. A very convenient arrangement for the regulation of the light is to have an Argand gas-burner with a switch, to which is attached, by a wire link, a brass or iron rod made of stout wire, which passes through the front of the case, terminating on the outside in a knob; when the polariscope is to be used for a series of observations, the gas-cock may be turned on full, and then the flame regulated, according to the requirements of the work, by means of the rod attached to the switch of the burner, by pulling it in or out according to the size flame desired.

The screw-caps of the observation-tube must not press too strongly upon the glass plates, as glass submitted to strains or pressure becomes capable of polarizing; rubber washers should be interposed between caps and plates.

WILD'S POLARISTROBOMETER.

This instrument was invented in 1864 by H. Wild.* A

Fig. 17.

striking peculiarity of it is, that between the polarizing and analyzing Nichols, of which the first rotates, is interposed a *Savart's polariscope*, by which a number of black bands of interference are produced, which disappear for a known position of the polarizer; this position, which can be determined with great precision, forms the stopping-point (*merkmal*) for the operator. The light used is that of the sodium flame.

Description of the Instrument.—Two views are given in Figs. 17 and 18; the capital letters in one correspond to the small letters of the other. Upon a metallic standard, X, Fig. 18, is carried a brass frame, Y, at either end of which are the polarizing and analyzing arrangements; the light enters at *b*, Fig. 17, through a round diaphragm, *c*, and

* H. Wild, *Ueber ein neues Polaristrobometer.* Berne, 1865.

passes to the Nichol prism *d*, which is joined to the scale *e*, and turns with it. The polarized ray passes through the observation-tube and arrives at the ocular of the polariscope. This part of the apparatus produces the phenomena of interference, and consists of two plates, *g*, of calc-spar three millimetres thick, cut at an angle of 45° to their optical

Fig. 18.

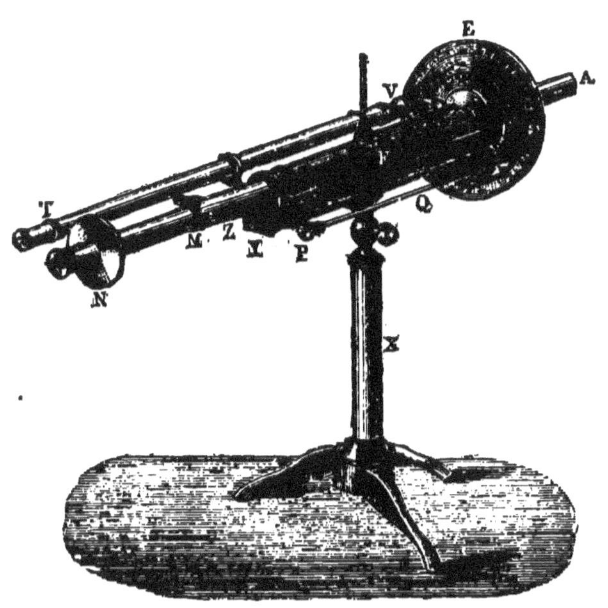

axes and cemented together again, so that their principal sections are at right angles to each other; there is a small telescope, magnifying about five times, whose lenses are shown at *h* and *i*. Between these, and in the focus of *h*, is a round diaphragm four millimetres in diameter and containing cross-wires. Finally the analyzing Nichol *l*, which is fixed, has its principal section horizontal; with the latter the crossed principal sections of the plate *g* must form an angle of 45°. At *m m* is a wide slit, which by the screws may be altered in size, serving to adjust the zero-point of the instrument. In order that the relative position of the parts *g* and *l* should remain unchanged, the ocular-tube, in which is contained the Nichol prism and the lens, is fastened by a pin and a slot, as shown in the Fig. 18. The whole polariscope is contained in the tube Z, arranged so that it can rotate through a small angle; N is a shield to protect the eye from the light. In the rotation of the polarizing prism, the brass plate on which is engraved the scale rotates also; this movement is effected by the rod P Q. The fixed index, *r*, serves to show the amount of the rotation. For reading the divisions of the scale the telescope *s* is provided, at the end of which is an opening, V, with a mirror which throws the light of a small gas flame on the scale. The source of light is the sodium flame, consisting of a Bunsen burner or alcohol lamp in which is kept a small globule of chloride of sodium fused on a platinum wire. Laurent's monochromatic lamp is an excellent arrangement for producing the sodium flame, and may be used for any saccharimeter requiring that kind of illumination. It consists (Fig. 19½) of a vertical Bunsen

Fig. 19½.

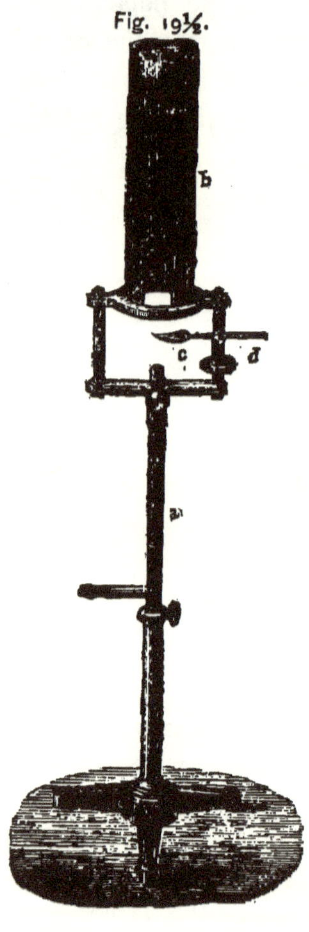

burner, *a*, surmounted by a chimney, *b*; *d* rotates and carries a fine platinum wire on which is fused some sodium chloride or carbonate, *c*.

The Use of the Apparatus.—For the execution of an observation the empty tube is placed in the apparatus, and the ocular of the polariscope, by the screws *m m*, is opened wide enough to admit of a clear view of the cross-wires. Turning the polarizer by means of *p*, Fig. 17, we find such a position that the illuminated field shows a number of parallel black lines or *fringes*, as in Fig. 19, *a*.

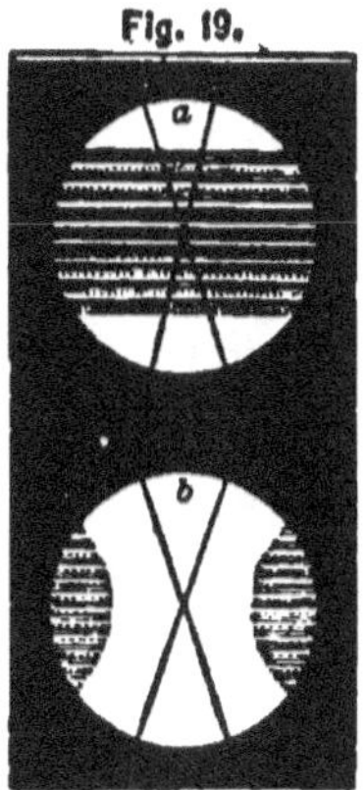
Fig. 19.

By continuing the rotation there arrives a time when a clear portion, free from fringes, appears on the field, and we can, by moving the button *p* to and fro, distribute the fringe-free portion symmetrically on the field with reference to the cross-wires, as shown by Fig. 19, *b*. This appearance serves as a stopping-point for the operation, and the reading of the scale should be 0° if all adjustments are correct. If the polarizing Nichol be turned still further, the fringes again increase in intensity, and finally become faint once more, the field presenting the same appearance at 90°, 180°, and 270° as it did at zero. The reading may be made at all of these points, and the results should agree. The disappearance of the fringes corresponds to a position of the rotating prism, when its principal section coincides with, or is at right angles to, that of the first plate of the calc-spar prism *g*. The greatest intensity of the fringes is observed when they are inclined at an angle of 45°.

If, after the zero-point of the scale has been sufficiently

verified, the observation-tube is filled with an optically active solution, the fringes of interference appear again, and the polarizer is then turned until, after several trials, the field presents the appearance shown in Fig. 10, *b*. When this point is attained the rotation is suspended, and the reading corresponding to the amount of sugar in the solution is taken.

This description of the polaristrobometer has reference to the apparatus with a circular scale divided into degrees from 0 to 360. A sugar scale has been added by dividing this into four hundred equal parts.

To estimate the sugar in a saccharine product 20 grms. are weighed and dissolved to 100 c.c., or 10 grms. to 50 c.c., and the observation taken in the 200 mm. tube. The reading is to be halved to show percentages of sugar. Where the assay contains but a small amount of sugar forty or sixty grammes may be weighed, dissolved to 100 c.c., and the result divided by four or six, as the case may be.

SHADOW SACCHARIMETERS

(*Saccharimètre à pénombre*).

A distinguishing peculiarity of this class of saccharimeters is that for a certain position of the optical parts the field of the instrument appears divided into two halves, the one very bright and the other as dark. For another position the whole field assumes a uniform grayish shadow, without any trace of vertical line.

To Prof. Jellett, of Dublin, belongs the credit of first inventing an instrument of this kind, though it has been much improved by the labors of Duboscq and Cornu. The source of light is monochromatic.

DUBOSCQ'S SACCHARIMÈTRE À PÉNOMBRE.

Fig. 20 shows the apparatus devised by Duboscq and

Fig. 20.

Cornu. The polarizing prism is of peculiar construction. A rhomb of calc-spar is divided longitudinally, following the plane of the smaller diagonal A B, Fig. 21, and each of the cut faces are removed for an angle of two and a half degrees, the sections *i* A B and A B *o* being taken off; the remaining parts are cemented together again on the planes passing through B *i* and B *o*. A double prism is thus obtained, of which the principal sections are at an angle of 5°. Owing to this construction, for very small changes in the luminous field a comparatively large angular rotation of the analyzer is required, and hence the delicacy of the instrument is assured.

Fig. 21.

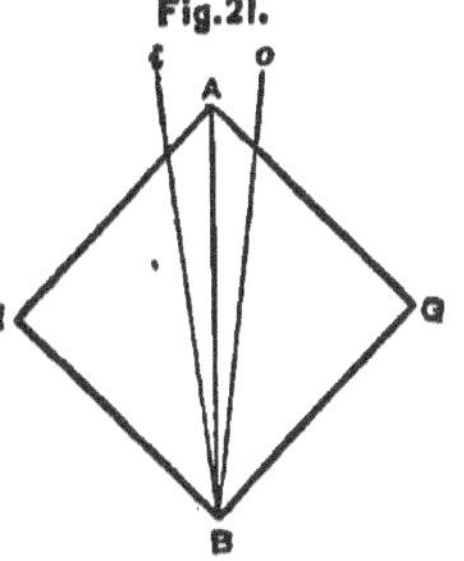

On filling the observation-tube with water, and placing the zero of the vernier to correspond to that of the scale, an observation through the ocular shows a vertical line separating two half-disks, which should appear of the same intensity. If they are not, the instrument is rectified by turning in one direction or the other the button O shown in the figure, which rotates a Nichol prism. When the whole surface of the field is of a uniform color, and the zero of the scale corresponds exactly to that of the vernier, the apparatus is properly adjusted, and is ready for the examination of sugar solutions. The normal weight (16.19 grms.), the amount of dilution, etc., are the same as in the case of the Soleil-Duboscq saccharimeter (page 135). When the observation-tube is filled with solution and placed in the saccharimeter, on viewing the luminous field through the ocular it will be seen that the equality of tone in the two half-disks no longer exists, one of the latter being much brighter than the other. The arm P is now slightly moved, and it is observed whether the inequality increases or diminishes. If the inequality increases, it is necessary to turn P in the opposite direction; if it diminishes, it may be made to disappear entirely by continuing the rotation of the arm. When the field assumes a uniform tinge, and the vertical line has entirely disappeared, the rotation is ceased and the scale read. The saccharimeter is provided with two scales on the circular plate, one indicating angular degrees, and the other percentages of sugar. A scale for milk-sugar and diabetic sugar is added in some instruments.

In both of Duboscq's saccharimeters a thickness of one millimetre of quartz corresponds to an angular rotation of 21.48°, which is also equal to that produced by a sugar solution containing 16.19 grms. of pure sugar in 100 c.c. The light used is monochromatic, and may be obtained by means of the Laurent lamp (Fig. 19½).

Duboscq's* *saccharimètre à pénombre* is much used in France, is very accurate, not expensive, and, with the improvements recently made upon it, is one of the most useful saccharimeters we have. It has the advantage, shared by all shadow saccharimeters, that persons who are color-blind are not necessarily prevented from working with it.

SCHMIDT AND HÄNSCH'S SHADOW SACCHARIMETER.

This instrument is of the same general form of the Soleil-Ventzke saccharimeter, though it differs materially in the optical portions. It makes use of the wedge-shaped quartz compensator and Jellett's prism (Fig. 21). Ordinary lamp-light, and not the monochromatic flame, is required. Stammer,† who has examined it, recommends the apparatus highly, not only for the sharpness and delicacy of its readings, but also for the facility with which colored solutions may be observed. It is provided with the ordinary observation-tubes of 200 mm. length, and also of 400 mm. and 600 mm. for the accurate testing of dilute sugar solutions. The readings show percentages of sugar, and not circular degrees.‡

LAURENT'S SACCHARIMETER.§

This apparatus differs materially from the preceding in

---

* Makers' address: J. & A. Duboscq, 21 Rue de l'Odéon, au fond de la cour, Paris.

† *Lehrbuch der Zuckerfabr.*, Ergänzungsband, 431. *Zeit. f. Rubenzucker*, 1880, 1098.

‡ Makers' address: Stallschreiber Strasse, No. 4, Berlin.

§ Maker's address: L. Laurent, 21 Rue de l'Odéon, Paris.

its optical parts, though the phenomena incident to the different appearances of the field are similar.

Figures 22 and 23 show the construction: *a* (Fig. 22) is

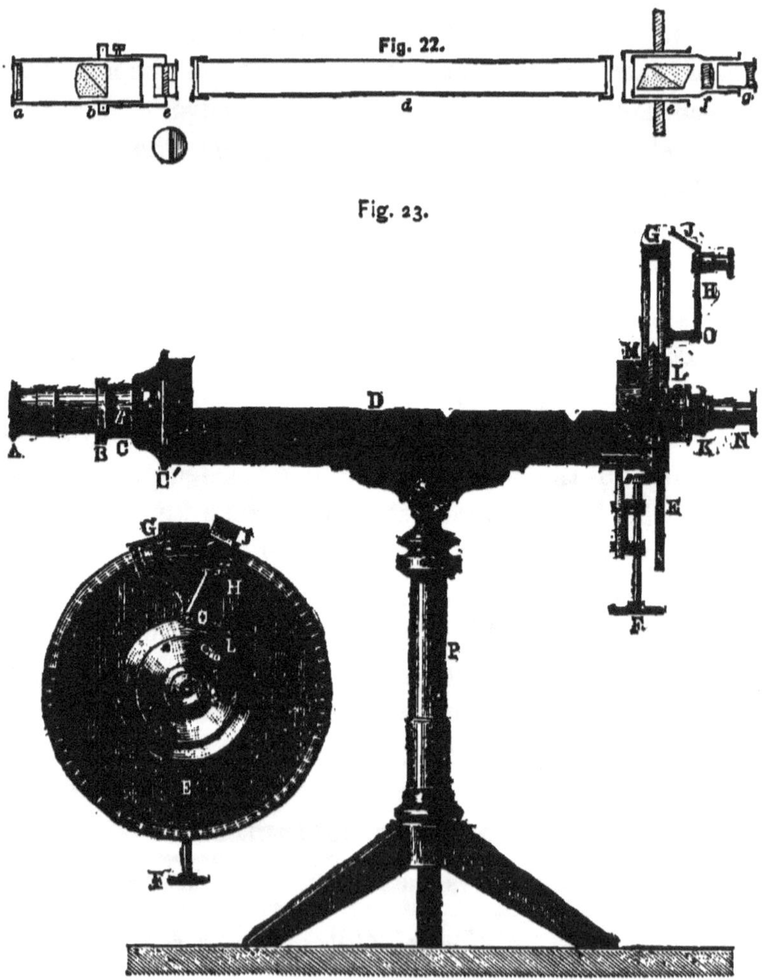

Fig. 22.

Fig. 23.

a thin plate of bichromate of potassium, which serves to cut off any blue or violet rays in the sodium light, thus rendering it more fully monochromatic; *b*, the polarizer, is a calc-spar prism. These two parts are placed in the movable brass tube A B (Fig. 23), which may be kept in any desired position by a set screw at β. c is a round diaphragm covered by a plate of glass, to which is cemented a thin section of quartz, cut parallel to its axis, in such a manner that only one-half of the aperture is covered by it; e is the analyzing Nichol, and f g the lenses of the telescope. The general arrangement of the instrument may be readily seen from the cuts.

The theory of the saccharimeter is as follows: If we suppose the plane of polarization to be vertical to the optical axis of the quartz plate, the light will traverse it without deviation; if the analyzer is rotated, we pass progressively to the maximum or total extinction of the light. Consequently, if we turn the analyzer through any given angle, a, to the right, the plane of polarization being no longer parallel to the axis of the crystal, the polarized ray will pass without deviation on the right side, on which there is no quartz; but on the left it will be deviated, and there will be determined on this side a principal section symmetrical to that of the polarizer on the right side, forming an angle equal to a, but to the left. If now we turn the analyzer until its principal section is perpendicular to that of the polarizer, there will be a total extinction of the light to the right, but only partial to the left. On the contrary, if the principal section of the analyzer is perpendicular to that which corresponds to the quartz plate, then there will be total extinction to the left and partial to the right. If, finally, the principal section of the analyzer is intermediate in position—that is, perpendicular to the axis of the crystal or horizontal—there will be partial extinction both to the right and left, and of equal intensity, and the luminous disk constituting the field of the instrument will appear uniformly in shadow. We can readily see from the foregoing that the smaller the angle *a*, the darker the shadow, and also that a small rotation of the analyzer tends to break the uniformity of the shadow; hence the saccharimeter is more sensitive when the angle *a* is less. With solutions much colored, by turning A B, Fig. 23, we augment the angle, by that means greatly brightening the field, thus enabling the operator to work with darker solutions than could be used otherwise. This is a considerable advantage, and forms a distinguishing peculiarity of the Laurent saccharimeter.

On looking through the ocular of the apparatus, and turning the analyzer until the medial line disappears and a uniform shadow is obtained, if the zero of the vernier does not exactly correspond to that of the scale, it may be made to do so by moving the screw L, Fig. 23. The apparatus figured only indicates circular degrees, but it is now made with a scale reading directly percentages of sugar. As with the Duboscq shadow saccharimeter, 16.19 grammes (the normal weight) of pure sugar in 100 c.c. is equivalent to an angular rotation of 21.48°, or 100 divisions of the scale, each corresponding to one per cent. of cane-sugar. The light used is that of the sodium flame (Fig. 19½).

The Laurent saccharimeter is a valuable instrument, and has been adopted for use in the French Government laboratories for the analysis of sugar. It has recently been improved so as to differ somewhat from the form above described, mainly in the direction of mechanical alterations, so as to work with a longer observation-tube; and in some other respects.

EQUIVALENCE IN DEGREES OF DIFFERENT SACCHARIMETERS.

1° Scale of Mitscherlich	= .750	grm.	sugar	in 100 c.c.
1° " Soleil-Duboscq	= .1619	"	"	"
1° " Ventzke-Soleil	= .26048	"	"	"
1° " Wild (sugar scale)	= .1000	"	"	"
1° " Shadow sacchar. (of Laurent and Duboscq)	= .1619	"	"	"

1° Mitscherlich	= 4.635° Soleil-Duboscq.
1° "	= 2.879° Soleil-Ventzke.
1° Soleil-Duboscq	= .215° Mitscherlich.
1° " "	= .620° Ventzke-Soleil.
1° " "	= 1.619° Wild.
1° " Ventzke	= .346° Mitscherlich.
1° " "	= 1.608° Soleil-Duboscq.
1° " "	= 2.648° Wild.
1° Wild (sugar scale)	= .618° Soleil-Duboscq.
1° " "	= .384° Soleil-Ventzke.
1° " "	= .133° Mitscherlich.

Equivalence in Circular Degrees.—

Wild (sugar scale)	1° = .1328	circular degree	D
Soleil-Duboscq	j 1° = .2167	" "	D
"	j 1° = .2450	" "	j
Soleil-Ventzke	j 1° = .3455	" "	D
"	j 1° = .3906	" "	j

Instruments reading angular degrees, such as Wild's, Laurent's, and Duboscq's saccharimètre à pénombre, may be made to give the concentration—*i.e.*, the number of grammes of sugar in 100 c.c. of solution—by the following formula:

$$c = \frac{100\,a}{k\,[a]\,D}$$

in which the observed angle of rotation is a, k the length of the observation-tube in decimetres, and $[a]$ D the specific rotatory power of cane-sugar, which for most purposes may be placed at 66.4°. When the specific gravity of the solution operated upon is known, the percentage by weight can be calculated by dividing the value of c obtained as above, by the density.

DECOLORIZING OF THE SUGAR SOLUTION.

Basic Lead Acetate.—The sugar solution to be tested in the optical saccharimeter is commonly more or less dark and requires to be decolorized. For this purpose the most ordinarily used and effective reagent is the solution of the *basic lead acetate.* It is prepared by boiling for half an hour, four hundred and forty grammes of neutral lead acetate with two hundred and sixty-four grammes lead oxide (litharge), and one and a half litres of water, and diluting when cool, to two litres; after standing some time the clear liquid may be siphoned off from the insoluble residue. The solution has a density of 1.267.

Alum.—Kohlrausch recommends the employment of alum solution in connection with the lead salt, which, by forming sulphate of lead, tends to more completely precipitate the coloring matter than when the acetate is used alone; sulphate of soda and other salts have also been suggested, though the chemical action is similar to that of alum. Woussen * adds a little tannin solution, before the addition of the lead salt, for very colored solutions.

Hydrate of Alumina.—Dr. Scheibler † prefers the use of precipitated hydrate of alumina dispersed in water as

* *De l'Analyse des Sucres*, 34.

† *Zeits. f. Rubenzuckerind. des Deut. Reiches*, 1870, 223.

a decolorizing agent, especially in highly-colored solutions. For the products of the beet this agent works well, but for low cane-sugars and molasses the decolorizing power is entirely insufficient when used alone. This reagent is prepared by precipitating a solution of alum with caustic ammonia in slight excess, and washing the resulting magma until the washings cease to render red litmus-paper blue. After the addition of the alumina to the sugar solution it should stand, with frequent shaking, for five or ten minutes before filtering.

Error from Use of Lead Solution.—There is one objection to the excessive use of lead which has not received the attention from sugar chemists that its importance merits—viz., the influence which the basic acetate of lead exerts upon invert-sugar in increasing its rotatory power. C. H. Gill * first pointed out this source of error in saccharimetric determinations, and explained the action by asserting that a compound of basic lead salt and levulose was formed. My own experiments † amply confirm his results. The tendency of the error is to increase with the amount of invert-sugar with the quantity of lead solution added; hence the error will be greatest in the darker-colored solutions, which generally not only contain a large portion of invert-sugar, but also require a proportionate amount of the clarifying liquid. For solutions poor in invert-sugar, and for which little lead solution is required, the error becomes very small and may be altogether neglected for all ordinary work. For solutions requiring more lead the least quantity should be added that will give a

---

* *Journ. Chem. Soc.*, April, 1871.

† I. *White refined sugar* free from invert-sugar polarized 90.3. After the

solution capable of being reliably tested in the saccharimeter.

Error from the Volume of Lead Precipitate.—The voluminous precipitate produced by lead in raw sugar solutions is itself a source of error. Scheibler* states that 100 c.c. beet-juice with 10 c.c. lead solution gave a precipitate whose volume was 1.3 c.c., which, by taking the place of the water in the flask, was equivalent to filling the latter to 98.7 c.c. instead of 100 c.c., introducing an error of .15 per cent. Nebel and Sostman† found for beet-juice the error to be .17 per cent., and with diffusion juices .27 per cent. Pellet‡ gives the following as the greatest error from this source: for beet-juice, .15 per cent. to .20 per cent.; cane-juice, .10 per cent.; *masse cuite*, .25 per cent.; second

addition of 9 per cent. by volume of solution of basic lead acetate it polarized 90.2.

II. *Centrifugal raw sugar* containing 2.74 per cent. invert-sugar, polarized 85.7 without lead.

With 2 per cent. lead solution it polarized 85.6.
With 5 " " " 85.7.
With 9 " " " 85.8.

III. *Muscovado sugar* containing 5 per cent. invert-sugar, polarized without lead 75.0.

With 6 per cent. lead solution, 75.8.
With 9 " " 75.9.

IV. *Low-grade refined sugar* containing 8 per cent. invert-sugar, polarized 77.0.

With 2 per cent. lead solution, 77.3.
With 5 " " 77.5.
With 9 " " 78.2.

V. *Sugar-house syrup* containing 28 per cent. invert-sugar, polarized 13.1.
With 4 per cent. lead solution, 13.3.
With 8 " " 13.7.

Another solution polarized 12.8.
With 10 per cent. lead solution, 13.7.

The lead solution itself has no action on the polarized ray.

* *Zeits. f. Zuckerind. des Deut. Reiches*, 1875, 1054.

† *Ibid.*, 1876, 724.

‡ *Ibid.*, 1876, 730.

and third product sugars (beet), .25 per cent.; molasses (beet), .63 per cent.

Scheibler* gives the following way of eliminating this error, which he calls the *method of double-dilution:* To 100 c.c. of the sugar solution 10 c.c. of lead solution are added and the saccharimetric reading taken. A second solution is prepared by mixing the same volumes of the saccharine liquid and lead solution, which is diluted to 220 c.c. and polarized. The last reading is doubled and subtracted from the first, the difference multiplied by 2.2, and this product taken from the first reading. This last result constitutes the corrected sugar content. Example:

A sugar solution gave a saccharimetric reading of 47.10.
After dilution.................................. 23.40.

(1) $23.40 \times 2 = 46.80$; $47.10 - 46.80 = .30$.

(2) $.30 \times 2.2 = .66$; $47.10 - .66 = 46.44$.

It often happens that, even after the addition of an excessive quantity of lead solution, the filtered liquid retains a strong brown color, rendering it unfitted to give an accurate reading. In this case a shorter observation-tube may be used or the solution made of half the normal strength. This mode of proceeding is, however, open to the objection that the necessary doubling of the reading increases the errors of observation.

Bone-Black.—This is the agent best suited to assist lead in the decolorization of raw sugar solutions. For this purpose a quantity of well-dried, powdered black should be kept on hand in a tight bottle with a wide mouth, fitted with a stopper that carries a glass tube the end of which is

* *Zeits. f. Zuckerind.*, 1875, 1054.

kept closed with a small cork when the bottle is not in use (Fig. 24). The bone-black should be dried at 120° for two hours.

Fig. 24.

After the addition of lead the flask is filled up to the mark with water, shaken, about one-half of the contents poured out, the black dusted into the liquid remaining in the flask, which is agitated vigorously a few moments and filtered. The least quantity of black should be used that will be sufficient to decolorize the solution.

An animal black of very superior decolorizing power may be prepared from bone-black in grains, such as is used in the sugar manufacture. To the char to be treated, about one-third to one-half more hydrochloric or nitric acid is added than is necessary to dissolve all the phosphate and carbonate of lime present, and the mixture is warmed several hours to promote solution (the more thorough the solution of the mineral matter the higher the decolorizing power of the carbon will be). It is then thoroughly washed with boiling water until the washings cease to redden litmus-paper. After the washing is complete the carbon is dried at 120° and finely powdered for use, though the grains may be used for filtering in tubes. Bone-black well prepared according to this method has a much greater decolorizing effect than the ordinary, and hence a small quantity may be taken for a test. The most obstinate solution may be readily decolorized by its aid with a moderate quantity of lead solution.

Absorption of Sugar by Bone-Black.—Animal charcoal has the property of absorbing sugar from its solutions.

Scheibler* has found that 5.5 grms. dried black shaken with 50 c.c. of a solution containing 13.024 grms. of cane-sugar, renders the polarization .4 per cent. to .5 per cent. too low. J. M. Merrick † has, by a series of experiments, fully proven the existence of this source of error in the saccharimetric determination. His results agree with those of Scheibler. When the sugar solution is filtered through a column of black, the first third should be rejected and the test made on the remainder of the filtrate, which may be confidently relied upon to contain the normal amount of sugar. The error arising from the absorption of sugar by the bone-black may be corrected by determining the absorption coefficient of the dried black. It is good practice to use both bone-black and lead on all low-grade solutions, but in moderate quantity, as the errors tend to counterbalance each other. For 50 c.c. of the Ventzke sugar normal solution with the specially prepared black, there may be used from ½ per cent. to 3 per cent. by volume of lead solution, and from ½ grm. to 1 grm. of char for all products of moderate difficulty. For molasses and the most troublesome cases it will be found that very rarely will more than 5 per cent. of lead solution and 1 to 2 grms. of char be necessary. When the bone-black is used in the quantities indicated above no correction need be made for its absorbing power for sugar. Even when working with ordinary bone-black two grammes are generally sufficient for all but the worst cases.‡

---

* *Zeits. f. Zuckerind. des Deut. Reiches*, 1870, 218.

† *Chem. News*, xxxviii. 33.

‡ H. A. Mott (*Journ. Am. Chem. Soc.*, i. 12), working with the Ventzke normal solution, has found that ten grammes of char absorb the following quantities of sugar:

PREPARATION OF PURE SUGAR.

The purest loaf-sugar of commerce is reduced to a fine powder, placed in a funnel whose barrel is stopped with a plug of raw cotton or sponge, and 85 per cent. alcohol poured upon it. This is allowed to percolate through the mass until a volume of alcohol has passed through equivalent in bulk to about three times that of the sugar. The latter is then drained, air-dried, powdered again, and finally dried in small quantities, at a time when it is needed, at a water-bath heat for half an hour. Sugar thus prepared may be confidently relied upon to indicate one hundred per cent. of cane-sugar with a correct saccharimeter.

ERRORS INHERENT IN THE OPTICAL METHOD OF ESTIMATING CANE-SUGAR.

1. **Influence of Temperature.**—Though the specific rotatory power of cane-sugar is not dependent in any marked degree on the temperature at which the observation is taken, yet the temperature has some effect, owing to a variety of causes, among which are (1) the alteration in the length of the observation-tube by changes of heat, (2) the increase in volume of the sugar solution in the tube and consequent change in density, and (3) the expansions and contractions produced in the quartz plates and other parts of the apparatus. Mategczek * gives a table of the corrections to be made at various temperatures for the Soleil-Ventzke and the Soleil-Duboscq instruments, 17½° C. being taken as a standard :

With pure sugar, .30 per cent. to .35 per cent.
" raw sugars, .10 " .66 "

Also, that some bone-blacks absorb in a different ratio from others.

* *Zeits. f. Zuckerind. des Deut. Reiches*, 1875, 877, 891.

	Soleil-Ventzke.		Soleil-Duboscq.	
Temp.	Reading at given temp.	II.	Reading at given temp.	II.
10°	100.17	26.004	....	....
11	100.14	26.010	....	....
12	100.12	26.016	....	....
13	100.10	26.022	....	....
14	100.08	26.028	....	....
15	100.05	26.034	100.05	16.181
16	100.03	26.039	100.03	16.184
17	100.01	26.045	100.01	16.188
17½	100.00	26.048	100.00	16.190
18	99.99	26.051	99.99	16.192
19	99.96	26.057	99.96	16.196
20	99.94	26.064	99.94	16.200
21	99.91	26.071	99.92	16.203
22	99.88	26.078	99.89	16.207
23	99.85	26.086	99.87	16.211
24	99.83	26.093	99.85	16.215
25	99.80	26.100	99.82	16.218
26	99.77	26.108	....	....
27	99.74	26.116	....	....
28	99.71	26.124	....	....
29	99.68	26.132	....	....
30	99.65	26.139	....	....

The numbers in the second column indicate the quantities to be weighed at corresponding temperatures to give a correct reading at 17½° C. If the correction is taken from the first column, the ordinary normal quantity must be weighed.

For saccharimeters reading circular degrees, the correction is made by adding the product of the difference in temperature from the normal 17° C. and the factor .011, to the degree read, when the temperature is above 17°, or subtracting when it is below. As, for example, the reading is 25° at 20° C.; then $20 - 17 = 3 \times .011 = .033$, which, added to 25°, makes 25.033° as the corrected result. This is for solutions of 25 grms. to 100 c.c., approximately.

II. **Personal Error.**—In all saccharimeters where the

reading is taken by a comparison of the equality of tint of two half-disks, there is a small but pretty constant source of inaccuracy in the results, owing to the fact that all eyes are not equally sensitive to minute differences of color. The same observer at different times of the day and in different conditions of the eye, from its being more or less fatigued, will give varied readings. Some persons are specifically unfitted for work with the polariscope having a field of two colors, on account of color-blindness; but this is not true of the shadow saccharimeters or that of Wild.

Dr. Landolt * has made a careful determination of this error with the aid of five experienced polarizers, and finds it to be, with the Soleil-Ventzke, the Soleil-Duboscq, and the Wild saccharimeters, from .3° per cent. to .5° per cent., plus or minus. Probably this will be considered too high; ±.2° per cent. would better represent the average. The personal error does not necessarily affect the ultimate accuracy of the results, for each operator can set the scale of the instrument to suit his own eye; and if more than one use the same instrument, each can have his personal correction. Thus, if one operator reads the zero-point at −.4° and another at zero, the former will have to add .4° to all of his readings.†

---

* *American Chemist*, iv. 18–20.

† Tollens (*Ber. Deut. Chem. Ges.*, 1877, 1403) and Schmitz (*ibid.*, 1877, 1414) have proved that the sp. rotatory power of cane-sugar is not constant for solutions of all concentrations. The effect on the results of the optical estimation of cane-sugar is too small to be taken into account for technical work, being less than one-tenth of one per cent. in all instruments. Elaborate tables of the correction for the different instruments have been calculated, and may be found in the places cited, and also in Stammer's *Lehrbuch der Zuckerfabrikation* and Landolt's *Optische Drehungsvermögen*.

ERROR OWING TO PRESENCE OF INVERT-SUGAR—OPTICAL INACTIVITY OF INVERT-SUGAR.

All raw sugars and molasses from the cane contain invert-sugar, sometimes in large amounts. Inasmuch as the rotatory power of invert-sugar made by acting upon cane-sugar with acids, is strongly to the left, a mixture of the two will give in the saccharimeter a reading too low as expressing the cane-sugar. It has been considered that the invert-sugar in the products of the cane possesses the same rotatory power as that artificially prepared, and it was customary with some chemists to correct their polariscopic readings by adding to them $\frac{34}{100}$ of the invert-sugar as found by the estimation with copper liquor.

Recent researches of Girard and Laborde* tend to verify a previous observation of Dubrunfaut—that the invert-sugar in cane products *is optically inactive*, and hence the use of the coefficient $\frac{34}{100}$ involves an error. The results of the above chemists are based on the examination of (1) syrups artificially prepared from raw cane-sugars of many sources ; (2) of molasses from the sugar plantations ; and (3) from the refiner's molasses. The cane-sugar was estimated directly with the polariscope, and also very carefully by inversion and gravimetric determination with copper liquor. The invert-sugar was determined in the same manner. In the majority of the samples examined the percentage of sugar by the copper method agreed quite closely with that by direct polarization, the latter being as often above as below the former.

Other investigators, among whom may be mentioned Muntz (*J. des Fabricants*, xvii. No. 5), Morin (*Sucrerie*

---

* *Journ. des Fabricants*, xvii. No. 5.

Indigène, xii. 158), Gill (*Sugar-Cane*, July, 1878), and Halse, have confirmed the conclusions of the French chemists by an extended series of experiments upon raw sugars of all origins, cane-juice, molasses, etc. Morin shows that analyses of raw sugar corrected by the coefficient $\frac{34}{100}$ generally add up over 100, even when large amounts of undetermined organic matters are present. As beet sugars and syrups rarely contain more than traces of invert-sugar, these results have no special application in that direction.

Meissl * has recently, by a most elaborate investigation, gone over the ground covered by the authorities named, with the result of completely contradicting both their facts and conclusions. Working with seven low-grade raw sugars from the cane, carrying from 5 to 13 per cent. invert-sugar, and determining the cane-sugar after inversion, and the invert-sugar directly, by Soxhlet's improved manipulation with Fehling's solution (page 201), he finds the cane-sugar by inversion to be always considerably higher than the saccharimetric reading, the difference varying with the amount of invert-sugar present; that by the use of the coefficient $\frac{34}{100}$ the corrected percentage of sugar agrees closely with that by inversion; that the syrups extracted from these sugars by alcohol, and containing from 27 to 39 per cent. invert-sugar, give essentially the same results as above. He also proves that the sugars, on complete analysis, do not add up over 100, but the quantity of organic matters not sugar, varies from 1 to 8 per cent. Meissl considers the conclusion of the chemists cited, as to the optical inactivity of invert-sugar in commercial products, as erroneous, and ascribes the error to the use of the gravimetric

* *Zeits. f. Rubenz.*, xxix. 1034; *Stammer's Jahresb.* (abstract), xix. 178.

method with Fehling's solution, which, he claims, gives results that are too high (page 203).

The coefficient $\frac{34}{100}$ is inadmissible, however, for general commercial work, because sugars are bought and sold (at least in the United States) on the direct polarization, and it would be clearly wrong to make the correction unless the matter was so understood by the merchant.

See also Horsin-Deon.*

INFLUENCE OF VARIOUS BODIES ON THE POLARISCOPIC READINGS.

Alcohol.—The presence of alcohol in solutions of cane-sugar does not alter materially the specific rotatory power; it diminishes the rotatory power of invert-sugar (Jodin—see Invert-Sugar, page 89).

Alkalies.—Caustic soda, ammonia, and potash lower the saccharimetric titre, according to Sostman,† and the effect may be represented quantitatively as follows:

Alkali in 100 c.c.	Strength of solution in sugar.		
	5 grms. in 100 c.c.	10 grms. in 100 c.c.	20 grms. in 100 c.c.
1 grm. K_2O	.426 per cent.	.65 per cent.	.915 per cent.
1 " Na_2O	.450 "	.907 "	1.217 "

* *Jr. Fabr. Sucre*, xx. No. 37.

† Sostman, *Zeits. f. Zuckerind. des Deut. Reiches*, 1866, 272.

Pellet's * results are somewhat different:

	5.4 grms. sugar in 100 c.c.	17.3 grms. sugar in 100 c.c.
1 grm. KOH	.17	.500
1 " NaOH	.14	.450
1 " NH_4O	.073	.085

Caustic lime has an important influence in lowering the specific rotatory power of cane-sugar. Muntz † gives the following in this relation:

Sugar solution, 10 grammes in 100 c.c.:

.409	gramme	sugar to	$\frac{1}{4}$	molecule CaO,	$[\alpha]$	D	64.9°
.818	"	"	$\frac{1}{2}$	"	"	"	61.3°
1.637	"	"	1	"	"	"	56.9°
3.274	"	"	2	"	"	"	51.8°

Pure cane-sugar being........................67.0°

In the estimation of cane-sugar, according to various observers, one part of lime lowers the rotation equivalent to—

.64	part of	sugar	(Jodin).
.79	"	"	(Dubrunfaut).
1.12	"	"	(Bodenbender).
1.22	"	"	(Stammer).

Baryta and *strontia* have a similar action to that of lime. On neutralization of the alkali or alkaline earth with acetic or phosphoric acids the normal rotation is restored.

Mineral Salts.—Muntz ‡ has found that some salts

* Pellet, *Zeits. f. Zuckerind. des Deut. Reiches*, 1877, 1036.

† Muntz, *ibid.*, 1876, 736.

‡ Muntz, *ibid.*, 1876, 735.

lower the specific rotatory power of cane-sugar. Taking the rotatory power of sugar at $[\alpha]$ D = 67.0°, he finds, in the case of chloride of sodium:

NaCl added.	Concentration of sugar solution in 100 c.c.		
	5 grms.	10 grms.	20 grms.
5 grms.	$[\alpha]$ D 66.1	66.2	66.3
10 "	65.3	65.3	65.6
20 "	63.8	63.7	61.0

Carbonates of soda, ammonia, and *potash,* and *phosphate of soda* have a small effect, 1 gramme of the salts in the sugar normal solution altering the rotation generally much less than .20 per cent. According to Bardy and Riche,* *sulphate, nitrate, chloride,* and *carbonate of potassium,* and *chloride of sodium* have little or no effect on the polarization. Muntz states that *sulphates of potassium, sodium, ammonium,* and *magnesium,* the *nitrates* and *acetates* of the same bases, *phosphate of soda, chlorates, sulphides, hyposulphides,* and *chlorides of calcium, magnesium,* and *barium,* alter the reading from 2 to 3 per cent. when they are present dissolved in the proportion of 20 to 30 parts to 100 parts of sugar.

CORRECTION OF THE MEASURING APPARATUS.

The graduated apparatus, as bought from the dealers, is seldom accurate, and requires to be corrected. For this purpose it is best to make standard flasks of 100 c.c. and 50 c.c. capacity, from which pipettes and all other measuring apparatus may be adjusted; the standards should be

* *Sucrerie Indigène,* x. 551.

kept in a safe place and only used for purposes of comparison. Flasks should be selected that will hold the required quantity of liquid up to a point a little below the middle of the neck which should not be too short. Clean and thoroughly dry the flask, place it on the pan of a balance in a room whose temperature is about 16° C., and counterpoise with weights; now, in the case of the 100 c.c. flask, weigh 99.89 grammes of distilled water, or, for the 50 c.c. flask, 49.945 grammes at 16° C., carefully wiping away any drops that may adhere to the neck. When the weighing is completed, mark on the neck of the flask a straight line tangent to the lowest curve of the meniscus formed by the surface of water. The weight of water taken is exactly equal to 100 grammes, or 50 grammes distilled water at 4° C., the temperature of water's greatest density. If the flasks are to be marked for two graduations, as 100 c.c. and 110 c.c. in one case, and 50 to 55 c.c. in the other, ten and five grammes of water respectively must be weighed after the 100 and 50 marks are fixed, and another mark made on the neck as before.

From the standard flasks standard pipettes, capable of exactly delivering 100 c.c. and 50 c.c., may be readily made by careful measurement with water, the mark placed on the pipettes indicating the exact volumes they will deliver into the standard flasks. By means of the pipettes the flasks for general use in the laboratory may be corrected; for this latter graduation no especial temperature of the water used is required, so long as it does not materially change during the progress of the correction. In this manner it is always easy, in a few moments, to graduate a flask with perfect accuracy—and in case of doubt the standard is always at hand.

CHAPTER VII.

Determination of Cane-Sugar—Chemical Methods.

Many of these methods have only an historical interest, and such will be but outlined in description; those, however, that are in actual use will be described in as much detail as the necessities of each case demand and the space will permit.

METHOD OF PELIGOT.

This process is based on the fact that lime enters into combination with cane-sugar in definite proportion, forming a sucrate, so that when excess of caustic lime is added to a sugar solution, an acidimetric estimation of the combined lime gives indirectly the amount of sugar dissolved. The method is only suited for saccharine products containing no grape-sugar, and it cannot be recommended for accuracy. It is executed as follows: Dissolve ten grammes of the sugar to be tested in 75 c.c. of water, add ten grammes of finely-powdered caustic lime to the solution, and agitate from seven to ten minutes, or until the lime is all combined with the sugar; for weak saccharine liquids a less quantity of lime may be used. Throw the milky liquid on a filter, and take 10 c.c. of the clear filtrate, dilute to 300 c.c., add a few drops of litmus solution, and titre with standard acid until the red color of the litmus just appears. The standard acid solution is made by dissolving twenty-one grammes of monohydrated sulphuric acid to a litre with water, that amount of solution being capable of saturating the lime which combines with fifty grammes of sugar; hence, 1 c.c. of the acid solution is equivalent to .05 gramme cane-sugar.

METHOD OF EXTRACTION BY ALCOHOL.

This is a method more particularly suited to the estimation of the sugar in plants, and where the quantities to be assayed are very small; when the conditions are favorable it is capable of giving accurate results. It consists in simply extracting the material with cold alcohol of specific gravity .830, and evaporating the alcoholic liquid obtained to dryness. Aqueous alcohol will dissolve small portions of invert-sugar, mineral salts, fat, and coloring matter, but these can be washed from the dried residue by means of absolute alcohol, which does not dissolve the cane-sugar. The process may be conducted as follows:* 100 to 120 grammes of the dried and finely-powdered substance are treated in a small flask with alcohol of .830 specific gravity, and a drop or two of a very dilute solution of caustic alkali is added to neutralize any acidity, the alcohol being allowed to stand in contact with the material under examination, with frequent shaking, for three hours; filter, add a fresh portion of alcohol, allow to stand two hours with agitation, and filter again. Repeat this operation several times, if necessary, as long as anything is taken up by the solvent. Unite the filtrates and evaporate at a gentle heat until a dry mass is obtained. Lastly, wash the residue repeatedly with absolute alcohol and dry in water-bath until it ceases to lose weight; the residue is calculated as pure cane-sugar. (See also Scheibler's method, page 266.)

* "Report on the Growth of the Beet in Ireland," *British Blue-Book.*

METHOD BY FERMENTATION.

The use of this process is open to many objections both from a want of exactness which is inherent in it, but also from the length of time required for its execution. One source of inaccuracy is that the fermentation does not always give the quantities of alcohol and carbonic acid in the normal proportions; sometimes a secondary fermentation takes place, with the formation of lactic acid and other bodies from which carbonic acid gas is not evolved.

The process may be carried out in two ways—viz., I. *By the estimation of the alcohol formed*, and II. *By the estimation of the carbonic acid.*

I. **By Estimation of Alcohol.**—When a solution of cane-sugar ferments, according to the best authorities, 100 parts of the sugar give 51.11 parts by weight of alcohol. The determination is conducted as follows: A rather dilute solution of the sugar is placed in a flask, and dry yeast added in quantity from 4 to 5 per cent. of the liquid, and the whole exposed to a temperature of from 20° to 25° C. When the fermentation is finished, which is in from 24 to 36 hours for a moderate quantity of sugar, the solution is submitted to distillation, and the amount of alcohol contained in the distillate is determined in the usual way.

II. **By the Estimation of Carbonic Acid.**—The solution is placed in a flask whose cork has two perforations, one of which carries a small glass tube just passing through the stopper and closed at its outer end during the fermentation; the other carries a tube bent at a right angle and connected with a U-tube containing fragments of pumice-stone moistened with concentrated sulphuric acid, which in turn is joined to a second U-tube filled with chloride of calcium in lumps. The proper quantity of yeast is added to the flask, and the whole system, consisting of flask and tubes, is weighed; it is then allowed to remain at a temperature of 20° to 25° C. until fermentation has ceased. An aspirator is now applied to the chloride of calcium tube, the stopper removed from the glass tube, and a current of air drawn through the arrangement, the flask being meanwhile moderately heated to facilitate the disengagement of the gas. The U-tubes serve to dry the gas so that no water may escape during the aspiration. The apparatus is now reweighed, and the difference between the first and second weights shows the quantity of carbonic acid produced during the experiment. 100 parts of cane-sugar correspond to 48.89 parts of carbonic acid.

When invert or grape sugar is present they will have to be determined separately, and the amount, calculated into cane-sugar, subtracted from the result given by fermentation; 475 parts cane-sugar = 500 parts invert-sugar.

DETERMINATION OF CANE-SUGAR BY FEHLING'S METHOD AFTER INVERSION.

Acids have the property of converting cane into invert sugar in definite proportion, so that 19 parts of the former produce 20 parts of the latter.

Execution of the Test.—To determine the cane-sugar, 1.00 gramme of the substance, if of a high tenor in sugar, and a proportionately larger quantity if the amount of sugar is lower, is dissolved in about 100 c.c. of water in a half-litre flask, 3 c.c. of strong hydrochloric acid added, and the whole heated for twenty minutes on a water-bath to 70°; the liquid is then nearly neutralized with caustic or carbonated alkali. When the contents of the flask have cooled, the solution is made up to the mark and is then ready for testing. The method of estimating the invert-sugar formed is, according to Fehling, either with Soxhlet's modification (page 201) or after the gravimetric method (page 203). For work with any pretension to accuracy the simple titration is quite inadmissible.

Calculation.—

$$\begin{aligned}&CuO \times .4307\\&Cu \quad \times .5394 = \text{cane-sugar.}\\&\text{Invert-sugar} \times .950 = \text{cane-sugar.}\end{aligned}$$

When invert-sugar is also present in the solution of which the cane-sugar is to be determined by inversion, the former is first estimated as a separate operation, and then a portion of the original solution is inverted as directed above, and the total invert-sugar, including that formed from the cane-sugar, is determined with the copper liquor.

An example will indicate the calculation required. The amount of invert-sugar present, as found by the direct test with the copper liquor, is 12.00 per cent.; for inversion 1.00 gramme of the substance is dissolved to 500 c.c., and of this solution 36 c.c. are necessary to precipitate 12 c.c. of the copper liquor; then

36 × .002 = .072 gramme of substance containing .060 gramme invert-sugar, or 83.33 per cent.

83.33
12.00 less invert-sugar originally present,

———

71.33, which is the figure representing the invert-sugar derived by inversion from the cane-sugar.

$$19 : 20 :: x : 71.33 = 67.76 \text{ per cent. cane-sugar;}$$
$$\text{or, } 71.33 \times \tfrac{95}{100} = 67.76.$$

When the oxide or metallic copper is weighed, the calculation is entirely similar.

In some cases the heating of the sugar solution with strong mineral acid causes a slight decomposition of the invert-sugar, which is shown by the liquid assuming a brown color. To avoid this Brunner recommends oxalic acid as the inverting agent.

CHAPTER VIII.

DETERMINATION OF DEXTROSE AND INVERT-SUGAR.

Section I. Fehling's Method and its Modifications.

THE basis of this method is a qualitative reaction for the detection of dextrose in the presence of cane-sugar, discovered by Trommer, whose results are summed up as follows: (1) An alkaline solution of copper oxide, containing a fixed organic acid, as tartaric, has the oxide reduced to suboxide by dextrose, and cane-sugar under the same circumstances is not at all, or only slightly, affected; (2) cane-sugar, when inverted by acids, is converted into a mixture of dextrose and levulose, which acts toward the alkaline copper solution precisely as grape-sugar; (3) there is a definite relation between the amount of oxide reduced and the sugar. The reaction takes place slowly in the cold, and almost instantly at the boiling temperature. By the oxidation of grape-sugar formic, acetic, and oxalic acids are formed. According to Reichardt, gummic acid is also produced; but this is denied by Claus,* who, however, found oxymalonic acid.

Barreswill first took advantage of Trommer's reaction to make it the basis of a quantitative method for the rapid estimation of cane and grape sugar. The solution proposed by him, consisting largely of alkaline carbonates, was found difficult to keep on account of the deposition of

* *Zeits. für Chemie*, 1869, No. 5.

oxide of copper. It was improved by Fehling,* who has investigated the quantitative relations of the bodies taking part in Trommer's reaction. He found that one equivalent of anhydrous grape or invert sugar was capable of reducing the oxide corresponding to ten equivalents of crystallized cupric sulphate—as:

$$\frac{1 \text{ eq. dextrose}}{C_6H_{12}O_6 = 180} = \frac{10 \text{ eq. cupric sulphate}}{CuSO_4 + 5H_2O = 2494.}$$

This has been confirmed by Neubauer.†

There are two ways of proceeding in regard to the estimation of sugars by the Fehling process—one making use of all the refinements of more recent discovery and requiring a considerable amount of time, being adapted for cases where the greatest accuracy is required; and the other quickly and easily executed, but quite exact enough for many technical purposes. It is proposed to discuss the subject divided as indicated above.

Part I. The Method as Suited for Technical Work —Volumetric.

Some recent researches have thrown doubt upon the constancy of the relation between dextrose and the amount of copper oxide reduced by it from alkaline solution, which affects both the volumetric and gravimetric methods. A conformity with these results would necessitate an alteration in the mode of operating, considerably lengthening it. These considerations, however, do not affect the substantial

---

* *Ann. der Chem. Pharm.*, lxxii. 106.

† *Arch. der Pharm.*, [2] lxxi. 278. Soxhlet denies that the relation between the copper salt and glucose is fixed, but that it varies according to the circumstances under which the test is made, from 9.7 to 11.1 equivalents of copper oxide to one of grape-sugar. See results of Soxhlet and others, page 201.

value of Fehling's process as ordinarily carried out for the greater part of commercial work, such as the analysis of raw sugars, syrups, etc., when the tenor is not higher than 20 per cent. There are cases that occur in commercial practice where the greatest possible exactness and care is required, and for which no analytical refinement would be misplaced. The analyst, however, must form his own judgment as to the proper course to pursue under any given circumstances.

Fehling's Solution.—The formula for this solution is as follows :

34.64 grammes pure cryst. cupric sulphate, dissolved in 160 c.c. dist. water;

150 grammes neutral potassium tartrate, dissolved in 600 c.c. to 700 c.c. of soda lye sp. gr. 1.12 (equivalent to about 90 grammes of the dry salt).

The two solutions are mixed and made up with water to a volume of 1000 c.c. at 15°. Of this

10 c.c. is equivalent to .050 grm. dextrose or invert-sugar;
" " .0475 " cane-sugar.

Fehling's solution, unfortunately, is not very stable, depositing oxide of copper in the cold, and especially when heated or exposed to light.

Violette * and Monier each give a formula for a solution which is said to keep well, but doubtless that of the latter is less to be recommended on account of its strong alkalinity.

* *Dosage du sucre au moyen des liqueurs titrées.*

Violette's Solution.—

34.64 grammes pure cryst. copper sulphate.
187 " tartrate of soda and potash (Rochelle salt).
78 " caustic soda.

The copper salt is to be dissolved in 140 c.c. of distilled water, slowly added to a solution of the tartrate and caustic soda, and the whole made up to one litre at standard temperature.

10 c.c. = .050 gramme dextrose or invert-sugar.
" = .0475 " cane-sugar.

Monier's Solution.—

40 grammes pure cryst. copper sulphate.
3 " chloride of ammonium.
80 " acid tartrate of potash (cream of tartar).
130 " caustic soda.

The sulphate is dissolved in 160 c.c. of water, the ammonium salt added, the solution mixed with the other ingredients dissolved in 600 c.c. of distilled water, and the whole made up to one litre.

10 c.c. = .0577 gramme dextrose or invert-sugar.
" = .0548 " cane-sugar.

The ammonium chloride furnishes free ammonia, which acts as a solvent for oxide of copper, thus preventing its precipitation on standing.

The investigations of several chemists * seem to establish that long boiling of cane-sugar with a strongly alkaline solution containing copper oxide causes a reduction of the oxide in small quantity. But, however, if (1) *the solution*

---

* Loiseau, *Amer. Chemist*, iv. 201. Felz, *ibid.*, iv. 113 ; iii. 313. Possoz, *Journ. des Fabr. des Sucre*, xiv. 50.

is diluted sufficiently, (2) if *the reduction takes place quickly*, and (3) if *the copper liquor used has merely enough alkali to ensure its permanence, either in the cold or at a boiling heat*, the error from this source is too small to affect the results notably for any purpose to which the method can be suitably applied.

Possoz's Solution.—Possoz recommends a copper liquor which he claims has no action on cane-sugar when used according to the directions given:

40	grammes	pure crystallized copper sulphate.
300	"	tartrate of potash and soda.
29	"	caustic soda.
150	"	bicarbonate of soda.

The sulphate of copper is dissolved in 150 c.c. of water and the bicarbonate added. The other salts are made into a solution with 500 c.c. of water, the two solutions mixed, boiled for one hour, allowed to cool, and water added to make one litre. The resulting solution is allowed to stand six months before use.

10 c.c. = .0577 gramme dextrose.
" = .0548 " cane-sugar.

The sugar solution to be tested by this process should be of such a concentration that .100 gramme dextrose or invert-sugar precipitates the copper from 30 c.c. of the copper liquor. The estimation is made by heating to 70° C. with a measured excess of copper liquor, filtering, and determining the copper remaining in the filtrate by a suitable method; whence the amount reduced by the dextrose may be calculated.

The formula following gives a cupric liquor which will be

found to be perfectly permanent; used with the necessary precautions, its action on cane-sugar may be altogether disregarded for ordinary work. It is the same as Violette's, except that the proportion of alkali contained is somewhat altered:

34.64	grammes	pure cryst. copper sulphate.
180	"	tartrate of potash and soda.
70	"	caustic soda.

Dissolve the tartrate and soda in 600 c.c. of distilled water, and to this add the cupric salt in solution, in small quantities at a time, shaking after each addition; when a clear liquid is obtained it is allowed to cool and made up to one litre.

10 c.c. = .050 gramme dextrose or invert-sugar.
" = .0475 " cane-sugar.

Selection of Reagents.—The tartrate and caustic alkali used may be of the best commercial quality, the latter as free from carbonate as possible. In order to obtain a copper salt containing rigidly the theoretical amount of oxide, the following procedure may be adopted: Procure a thoroughly reliable article of chemically pure crystallized cupric sulphate, or make it by recrystallizing the commercial salt, and select the clear, well-formed crystals, rejecting those that are opaque and which generally consist of a more or less wet aggregate of fine crystalline material; carefully brush the selected pieces from all fine powder, and pulverize them, repeatedly pressing the powder between sheets of filter-paper to get rid of any adhering moisture. Preserve the salt thus prepared in a closely-stopped bottle until it is to be weighed out for use. If pure sulphate of copper cannot readily be obtained, 8.804 grammes pure metallic copper, precipitated by the battery or otherwise, is dissolved in nitric acid, and sulphuric acid, in quantity slightly more than that necessary to combine with the copper, is added; the mixture is evaporated to drive off the nitric acid, the free sulphuric acid neutralized with caustic soda, and the copper sulphate thus obtained is used in the preparation of the copper solution by the above formula; 8.804 grammes of copper is exactly the amount contained in 34.64 grammes of the crystallized sulphate. Care must be taken to thoroughly dry the precipitated copper, and at so low a temperature that oxidation will not take place. The copper solution should be kept in a blue glass bottle or one blackened on the outside.

Strength of Sugar Solution.—The amount of sugar to be determined varies greatly in different products submitted to the grape-sugar estimation; it is best to make the sugar solution of such dilution that from 25 c.c. to 50 c.c. will precipitate the copper from 10 c.c. of the copper liquor; the sugar solution should not be much stronger than this, as it then becomes difficult to hit the end point of the reaction with sufficient delicacy.

Calculation of Results—Glucose Normal.—It is convenient to establish a standard strength, for the sugar solution, of 5 grammes of the substance to be assayed to 100 c.c., and this may be called the *glucose normal solution.* From the varying amounts of grape-sugar contained in the material to be examined it becomes necessary to vary from the glucose normal by weighing out 10, 15, or 20 grammes to 100 c.c. of volume, when the solution becomes *double*, *triple*, or *quadruple normal;* or to weigh 5 grammes of assay, and dilute to 200 c.c., 300 c.c., or 500 c.c., when the solution is called *half*, *third*, or *fifth nor-*

mal. The calculation of the results may be greatly abridged in the following way: The reciprocal of the number of cubic centimetres required of the glucose normal solution to precipitate 10 c.c. of the copper liquor, multiplied by 100, is the direct percentage of dextrose or invert-sugar sought; for

10 c.c. copper liquor = .050 gramme grape-sugar,

and the normal glucose solution contains in 1 c.c. — .050 gramme of the substance. Suppose in an experiment 30 c.c. of sugar solution is required for 10 c.c. of copper solution; then

30 × .050 = 1.50 grammes of assay used, containing
.050 gramme grape-sugar,

$$\frac{.05}{1.50} = 3.33 \text{ per cent. grape-sugar;}$$

the reciprocal of 30 is .0333, which, by displacement of the decimal point, becomes 3.33. A table for thus calculating percentages is appended.

TABLE FOR CALCULATING THE PERCENTAGE OF GRAPE OR INVERT SUGAR WHEN THE NUMBER OF C.C. USED REFERS TO THE "GLUCOSE NORMAL SOLUTION."

(5 *grammes to* 100 *c.c.*)

No. c.c. used.	P. ct. Grape or Invert Sugar.	No. c.c. used.	P. ct. Grape or Invert Sugar.	No. c.c. used.	P. ct. Grape or Invert Sugar.	No. c.c. used.	P. ct. Grape or Invert Sugar.	No. c.c. used.	P. ct. Grape or Invert Sugar.	No. c.c. used.	P. ct. Grape or Invert Sugar.
1	100.0	72	1.39	143	.699	214	.467	285	.351	356	.2809
2	50.00	73	1.37	144	.694	215	.465	286	.350	357	.2801
3	33.33	74	1.35	145	.690	216	.463	287	.348	358	.2793
4	25.00	75	1.33	146	.685	217	.461	288	.347	359	.2785
5	20.00	76	1.32	147	.680	218	.459	289	.346	360	.2778
6	16.66	77	1.30	148	.676	219	.457	290	.345	361	.2770
7	14.29	78	1.28	149	.671	220	.454	291	.344	362	.2762
8	12.50	79	1.26	150	.667	221	.452	292	.342	363	.2755
9	11.11	80	1.25	151	.662	222	.450	293	.341	364	.2747
10	10.00	81	1.23	152	.658	223	.448	294	.340	365	.2740
11	9.09	82	1.22	153	.654	224	.446	295	.339	366	.2732
12	8.33	83	1.20	154	.649	225	.444	296	.338	367	.2725
13	7.69	84	1.19	155	.645	226	.442	297	.337	368	.2717
14	7.14	85	1.18	156	.641	227	.440	298	.335	369	.2710
15	6.67	86	1.16	157	.637	228	.438	299	.334	370	.2703
16	6.25	87	1.15	158	.633	229	.437	300	.333	371	.2695
17	5.83	88	1.14	159	.629	230	.435	301	.332	372	.2688
18	5.55	89	1.12	160	.625	231	.432	302	.331	373	.2681
19	5.26	90	1.11	161	.621	232	.431	303	.330	374	.2674
20	5.00	91	1.10	162	.617	233	.429	304	.329	375	.2667
21	4.76	92	1.09	163	.613	234	.427	305	.328	376	.2659
22	4.54	93	1.07	164	.610	235	.425	306	.327	377	.2652
23	4.35	94	1.06	165	.606	236	.424	307	.326	378	.2645
24	4.17	95	1.05	166	.602	237	.422	308	.325	379	.2638
25	4.00	96	1.04	167	.599	238	.420	309	.324	380	.2631
26	3.85	97	1.03	168	.595	239	.418	310	.322	381	.2625
27	3.70	98	1.02	169	.592	240	.417	311	.321	382	.2618
28	3.57	99	1.01	170	.588	241	.415	312	.320	383	.2611
29	3.45	100	1.00	171	.585	242	.413	313	.319	384	.2604
30	3.33	101	.99	172	.581	243	.411	314	.318	385	.2597
31	3.23	102	.98	173	.578	244	.409	315	.317	386	.2591
32	3.12	103	.97	174	.575	245	.408	316	.316	387	.2584
33	3.03	104	.96	175	.571	246	.406	317	.315	388	.2577
34	2.94	105	.95	176	.568	247	.405	318	.314	389	.2571
35	2.86	106	.94	177	.565	248	.403	319	.313	390	.2564
36	2.78	107	.93	178	.562	249	.402	320	.312	391	.2557
37	2.70	108	.92	179	.559	250	.400	321	.311	392	.2551
38	2.63	109	.917	180	.555	251	.398	322	.310	393	.2544
39	2.56	110	.909	181	.552	252	.397	323	.309	394	.2538
40	2.50	111	.900	182	.549	253	.395	324	.309	395	.2532
41	2.44	112	.893	183	.546	254	.394	325	.308	396	.2525
42	2.38	113	.885	184	.543	255	.392	326	.307	397	.2519
43	2.32	114	.877	185	.540	256	.391	327	.305	398	.2512
44	2.27	115	.869	186	.538	257	.389	328	.305	399	.2506
45	2.22	116	.862	187	.535	258	.388	329	.304	400	.2500
46	2.17	117	.855	188	.532	259	.386	330	.303	401	.2494
47	2.12	118	.847	189	.529	260	.385	331	.302	402	.2487
48	2.08	119	.840	190	.526	261	.383	332	.301	403	.2481
49	2.04	120	.833	191	.523	262	.381	333	.300	404	.2475
50	2.00	121	.826	192	.521	263	.380	334	.299	405	.2469
51	1.96	122	.820	193	.518	264	.379	335	.2985	406	.2463
52	1.92	123	.813	194	.515	265	.377	336	.2976	407	.2457
53	1.89	124	.806	195	.513	266	.376	337	.2967	408	.2451
54	1.85	125	.800	196	.510	267	.374	338	.2958	409	.2445
55	1.82	126	.794	197	.508	268	.373	339	.2950	410	.2439
56	1.79	127	.787	198	.505	269	.372	340	.2941	411	.2433
57	1.75	128	.781	199	.502	270	.370	341	.2932	412	.2427
58	1.72	129	.775	200	.500	271	.369	342	.2924	413	.2421
59	1.69	130	.769	201	.497	272	.368	343	.2915	414	.2415
60	1.67	131	.763	202	.495	273	.366	344	.2907	415	.2409
61	1.64	132	.757	203	.493	274	.365	345	.2898	416	.2406
62	1.61	133	.752	204	.490	275	.364	346	.2890	417	.2398
63	1.59	134	.746	205	.488	276	.362	347	.2882	418	.2392
64	1.56	135	.741	206	.485	277	.361	348	.2873	419	.2387
65	1.54	136	.735	207	.483	278	.360	349	.2865	420	.2381
66	1.51	137	.730	208	.481	279	.358	350	.2857	421	.2375
67	1.49	138	.725	209	.478	280	.357	351	.2849	422	.2370
68	1.47	139	.719	210	.476	281	.356	352	.2841	423	.2364
69	1.45	140	.714	211	.474	282	.355	353	.2833	424	.2359
70	1.43	141	.709	212	.472	283	.353	354	.2825	425	.2353
71	1.41	142	.704	213	.469	284	.352	355	.2817	426	.2347

No. c.c. used.	P. ct. Grape or Invert Sugar.	No. c.c. used.	P. ct. Grape or Invert Sugar.	No. c.c. used.	P. ct. Grape or Invert Sugar.	No. c.c. used.	P. ct. Grape or Invert Sugar.	No. c.c. used.	P. ct. Grape or Invert Sugar.	No. c.c. used.	P. ct. Grape or Invert Sugar.
427	.2342	456	.2193	485	.2062	514	.1945	543	.1842	572	.1748
428	.2336	457	.2188	486	.2058	515	.1942	544	.1838	573	.1745
429	.2331	458	.2183	487	.2053	516	.1938	545	.1835	574	.1742
430	.2325	459	.2179	488	.2049	517	.1934	546	.1831	575	.1739
431	.2320	460	.2174	489	.2045	518	.1930	547	.1828	576	.1736
432	.2315	461	.2169	490	.2041	519	.1927	548	.1825	577	.1733
433	.2309	462	.2164	491	.2037	520	.1923	549	.1821	578	.1730
434	.2304	463	.2160	492	.2032	521	.1919	550	.1818	579	.1727
435	.2299	464	.2155	493	.2028	522	.1916	551	.1815	580	.1724
436	.2293	465	.2150	494	.2024	523	.1912	552	.1812	581	.1721
437	.2288	466	.2146	495	.2020	524	.1908	553	.1808	582	.1718
438	.2283	467	.2141	496	.2016	525	.1905	554	.1805	583	.1715
439	.2278	468	.2137	497	.2012	526	.1901	555	.1802	584	.1712
440	.2273	469	.2132	498	.2008	527	.1897	556	.1798	585	.1709
441	.2267	470	.2128	499	.2004	528	.1894	557	.1795	586	.1706
442	.2262	471	.2123	500	.2000	529	.1890	558	.1792	587	.1703
443	.2257	472	.2119	501	.1996	530	.1887	559	.1789	588	.1700
444	.2252	473	.2114	502	.1992	531	.1883	560	.1786	589	.1698
445	.2247	474	.2110	503	.1988	532	.1880	561	.1782	590	.1695
446	.2242	475	.2105	504	.1984	533	.1876	562	.1779	591	.1692
447	.2237	476	.2101	505	.1980	534	.1873	563	.1776	592	.1689
448	.2232	477	.2096	506	.1976	535	.1869	564	.1773	593	.1686
449	.2227	478	.2092	507	.1972	536	.1865	565	.1770	594	.1683
450	.2222	479	.2088	508	.1968	537	.1862	566	.1767	595	.1681
451	.2217	480	.2083	509	.1965	538	.1859	567	.1764	596	.1678
452	.2212	481	.2079	510	.1961	539	.1855	568	.1760	597	.1675
453	.2207	482	.2075	511	.1957	540	.1852	569	.1757	598	.1672
454	.2203	483	.2070	512	.1953	541	.1848	570	.1754	599	.1659
455	.2198	484	.2066	513	.1949	542	.1845	571	.1751	600	.1667

The use of the table is quite simple. When the volume of sugar solution used is very small or fractional, by a change of the decimal point, making a whole number, a more exact figure may be obtained, the operator always knowing approximately the percentage of grape-sugar, so as to be able to set down the result correctly. Thus an experiment gives 4.1 c.c. of the glucose normal; looking in the table for the percentage opposite 41, we find it to be 2.44, which, by change of the decimal point, gives 24.40 per cent. as the true result. If 410 is taken instead of 41 a still more exact result is obtained—namely, 24.39 per cent. In cases where the strength of the sugar solution varies from the normal, the result of the test in c.c. may be reduced to the standard by multiplying by 2, 3, or 5 respectively when the half, third, or fifth normal solution is used; or by dividing by 2, 3, or 4 for the double, triple, or quadruple normal solution.

Examples: I. 5 grammes of a raw sugar were dissolved to 300 c.c., and on estimation it was found that 36 c.c. were required for 10 c.c. of copper liquor; then $\frac{36}{3} = 12$, and the corresponding percentage in the table is 8.33. II. 20 grammes of cane-juice made to 100 c.c. required 24.6 c.c.; hence $24.6 \times 4 = 98.4$. Calling this 98.5, we find from the table that

	98	corresponds to	1.02	per cent.
and	99	" "	1.01	"
The average is	98.5	" "	1.015	"

which is the percentage sought.

When the estimation of grape-sugar is made on the same sample from which the cane-sugar is determined with the saccharimeter, it saves a great deal of time to have a pipette graduated to deliver 19.21 c.c. when the Soleil-Ventzke saccharimeter is used, and 30.8 c.c. for those having the normal weight of 16.19 grammes. These pipettes measure precisely 5 grammes of the original substance. In this way, after sufficient of the filtered solution has been taken for polarizing, from the remaining portion (free from lead) the required quantity is taken out with the appropriate pipette to make the glucose normal solution or its multiples. By this method of proceeding one weighing suffices for two determinations, and a further advantage is that a better average sample is obtained by weighing 26 or 16 grammes, while the glucose and cane-sugar determinations are made from identically the same solution.

When the sugar solution is sufficiently colored to interfere with the copper test, it may be readily decolorized by shaking with a very small quantity of bone-black and filtering.

Organic matter other than sugar exerts a very slight, if any, influence upon the results of the invert-sugar estimation after Fehling * (see note at the end of the volume).

---

* It is entirely inadmissible to use a sugar solution containing lead in Fehling's method, as the results of experiments given below show. This fact was pointed out by H. C. Gill (*J. Chem. Soc.*, 1871, April), who also suggested the use of sulphurous acid as a precipitating agent. Experiments that I have made on the subject confirm Gill's results, and establish further that the precipitation of the lead must be very complete, as a mere trace seems to interfere as much with the test as a larger quantity. For this reason sulphates are unsuitable as precipitating agents, sulphate of lead being perceptibly soluble in water and sugar solutions. The use of sulphurous acid is, however, not open to this objection.

EFFECT OF LEAD SOLUTION.

I. *Refined sugar* containing 4.25 per cent. invert-sugar.

A sugar solution required for 10 c.c. copper liquor	35.5 c.c.
The same, with excess of lead solution, required	43.5 "
The same, with lead solution and excess of Na_2SO_4, required	43.5 "

II. *Invert-sugar from cane-sugar.*

A solution to precipitate 10 c.c. of copper liquor required	53.5 c.c.
The same, with a large quantity of Na_2SO_4, without lead, required	53.6 "
The same, with lead solution and excess of Na_2SO_4, required	90 "

III. *Molasses-sugar* containing 2 86 per cent. invert-sugar:

A solution required to precipitate 10 c.c. copper liquor,	50.7 c.c.
The same, with a large quantity of Na_2SO_4, required	50.2 "
The same, with lead solution and excess of Na_2SO_4, required	68 "

EFFECT OF SULPHUROUS ACID.

I. A sugar solution required for 10 c.c. copper solution	31.2 c.c.
The same, with 5 per cent. SO_2 solution	31.0 "
" " 20 " " "	31.5 "
" " 30 " " "	31.2 "
" " 50 " " "	31.3 "
II. A solution of sugar required for 10 c.c. copper liquor	30.7 c.c.
The same, with 1 per cent. lead solution	37.5 "
" " 1 " " " and excess of SO_2	32.0 "

Execution of the Test.—10 c.c. of the copper liquor is measured into a porcelain dish or casserole, diluted with its volume of water, and with a good flame quickly brought to a boil. The liquid should show no signs of precipitation by ebullition. The sugar solution is added from a burette as rapidly as possible without risk of running in an excess. The color changes from a deep, clear blue to a dull hue, and at the same time the suboxide begins to form. As the operation proceeds the red color begins to manifest itself, and the liquid assumes a bluish-violet tinge, in which the red constantly increases, the shades passing through bluish red, violet red, dark crimson, and finally to a full crimson, when the copper is just thrown down. The last color changes to a bright scarlet as soon as invert-sugar or grape-sugar is present in excess. The experienced operator can easily estimate by the color of the boiling solution how the operation is proceeding, the end point of the reaction being indicated by the full crimson color of the agitated mass, without tinge of scarlet, and by the shade of the supernatant liquid after the suboxide has settled out, which should be a clear pearl white, neither bluish nor yellowish.

In order that the end point may be determined with greater exactness, it is necessary to remove a little of the liquid in a small pipette, filter, acidify with acetic acid, and add a drop of a very dilute solution of potassium ferrocyanide, which will strike a brownish-red color as long

---

III.	A sugar solution required	30.7 c.c.
	With 2 c.c. lead solution	39.5 "
	" 2 " " " + excess SO_2	30.5 "
	" 1 " " "	37.0 "
	" 1 " " " + excess SO_2	33.5 "
	" 1 " " " + larger excess SO_2	31.0 "

as copper remains in solution. If the color is very faint, a practised operator can estimate the amount of sugar solu-

Fig. 25.

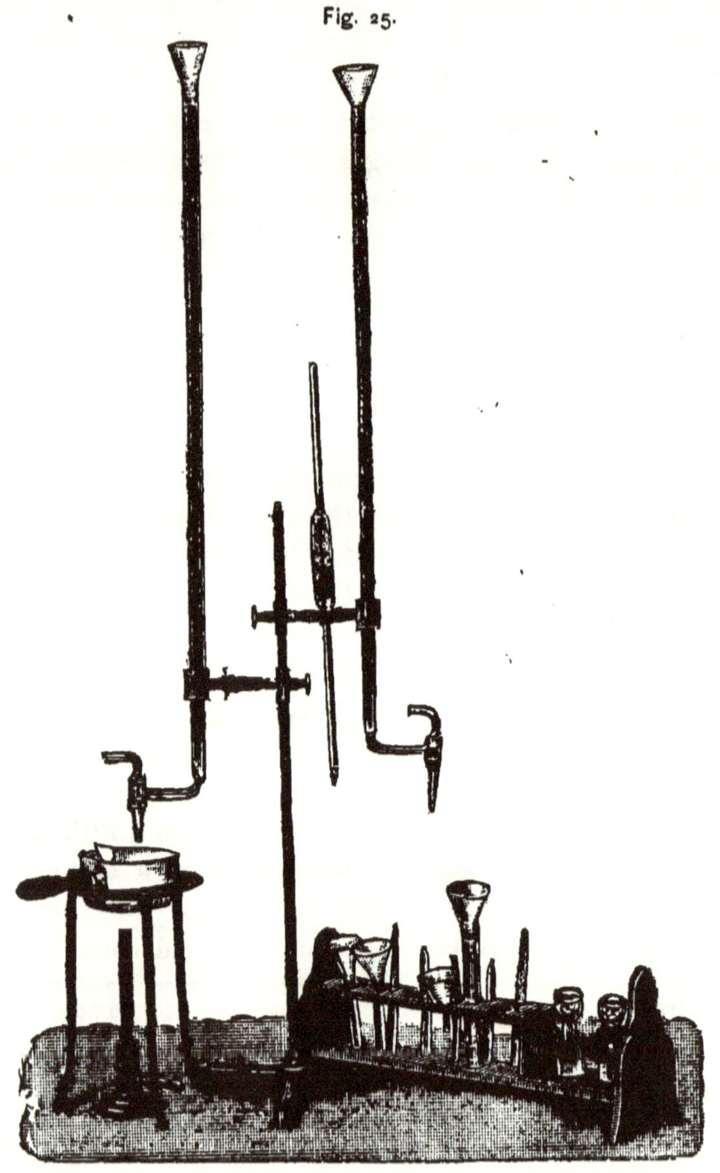

tion to be added, to that actually run in, to complete the reaction.

Apparatus.—Fig. 25 shows a collection of apparatus suitable for carrying out the method of testing described. The arrangement and choice of implements can be highly recommended where *rapid* work is required, especially when a number of tests are made at once. It consists of a burette-stand; two burettes provided with glass cocks, one of 50 c.c. capacity for the sugar solution, and a second of 100 c.c. for the copper liquor; the casserole over which the burette delivers the sugar solution; a pipette graduated to take out 5 grammes from the sugar normal solution of the polariscope; small pipettes (about four inches long) for obtaining a sample from the casserole; a rack for holding a number of 3-in. test-tubes, funnels ($\frac{3}{4}$ in.), and pipettes, together with the acetic acid and ferrocyanide of potassium. The latter are contained in dropping or atropia bottles, shown by Fig. 26. They have a piece of sheet gum stretched and tied over their tops, so that it is only necessary to press upon the gum in order to fill the tube with liquid.

Fig.26.

When reliance is placed in the disappearance of the blue color alone as the end point of the test, the operation may be performed in a large test-tube held over a sheet of white paper. The color of the liquid, after the precipitate has settled, is best seen by holding the tube in a slanting position. This method of carrying out the test is not so accurate as the one described, and *not* quicker in execution.

Pavy's Modification of the Volumetric Method.*

Dr. Pavy has taken advantage of the fact that caustic ammonia is a powerful solvent for suboxide of copper. In his process he uses a large excess of this reagent mixed with the ordinary copper liquor, so that the reduction takes place without precipitation, the boiling solution becoming decolorized at the same time. Under the conditions specified it was found that six molecules of cupric oxide are reduced by one molecule of dextrose, instead of five molecules as in the process of Fehling (old system). The copper liquor used by Pavy is of the following composition:

Cupric sulphate cryst.,	34.65	grammes;
Potassio-sodic tartrate,	173	"
Caustic potash,	160	"

in one litre. 120 c.c. of this solution is mixed with 300 c.c. of solution of ammonia sp. gr. .880, and water added to make one litre. 20 c.c. of this corresponds to .010 gramme of grape or invert sugar.

The test is made in a flask of about 80 c.c. capacity, with a cork fitted in the neck carrying a glass tube dipping under water, at the end of the tube being placed a piece of rubber tube cut longitudinally, so as to form a valve to prevent the water from being forced into the flask by condensation during a momentary stoppage of the operation. 40 c.c. of the dilute test liquor is run into the flask, and while boiling the sugar solution is added, drop by drop, until complete decolorization is effected. The results are said to be quite satisfactory. See original papers (*loc. cit.*)

---

* *Chem. News*, xxxix. 77, 197, 249; *ibid.*, Steiner, xl. 139.

Part II. The Method as Suited for Exact Work.

A. **Volumetric.**—*Soxhlet's Researches.*—As stated on page 186 (note), Soxhlet, by recent investigations, has demonstrated that the relation of dextrose and invert-sugar to the cupric oxide reduced from alkaline solution is not constant, contrary to the view formerly accepted, but varies within certain limits, according to the circumstances (mainly in regard to dilution) under which the reduction takes place. According to his results, 50 c.c. of a 1 per cent. dextrose solution, or 100 c.c. of a ½ per cent. solution, with the undiluted copper liquor, required for complete reduction from 101.0 c.c. to 101.4 c.c. Fehling's solution.

For the F. solution diluted with 1 vol. water, 99.5 c.c. F. sol.
" " " 2 " " 98.1 " "
" " " 3 " " 97.3 " "
" " " 4 " " 97.1 " "

In the case of invert-sugar there was required:

For the undiluted F. solution............ 101.2 c.c. F. sol.
For the F. solution diluted with 1 vol. water 99.5 " "
" " " 2 " " 98.2 " "
" " " 3 " " 97.3 " " *

These results correspond on the average, for the undiluted copper liquor, to the relation of

1 eq. sugar to 10.1 eq. CuO,

and for the fourfold diluted solution:

1 eq. sugar to 9.7 eq. CuO,

against the relation of

1 eq. sugar to 10 eq. CuO,

according to Fehling.

---

* In later experiments these results are varied from somewhat.

Execution of the Test.—Notwithstanding these facts, good results may be obtained, in the determination of invert or grape sugar, if precaution is taken to have the conditions in regard to dilution, and the relative amounts of copper solution and sugar, the same as indicated below, in which case the possible error is placed at ± .2 per cent. The copper reagent with Soxhlet is made of two solutions, viz.:

No. 1, consisting of 34.639 grammes cupric sulphate, dissolved in water to 500 c.c.; and

No. 2, of 173 grammes sodio-potassic tartrate (Rochelle salt), dissolved to 400 c.c., mixed with 100 c.c. soda-lye, containing 400 grammes to the litre.

These solutions are kept separately and mixed in equal volume when a test is to be made. To make an estimation, a preliminary experiment is tried to find the approximate strength of the sugar solution, and the latter is diluted to contain 1 per cent. of the sugar to be determined. 50 c.c. of this is heated to boiling for 2 to 4 minutes, with a quantity of the undiluted copper liquor judged to be nearly that necessary for reduction. The liquid mixture is then thrown on a filter, and the filtrate tested for copper with acetic acid and potassium ferrocyanide. If the metal is present a new experiment is made with the same quantity of sugar solution and less copper liquor, and so on, until in two consecutive experiments, differing by $\frac{2}{10}$ c.c. copper solution, the filtrate of one shows copper while the other does not. From the quantity of copper solution corresponding to the sugar the calculation can be made, 50 c.c. of the solution being equivalent to .500 gramme dextrose or invert-sugar.*

* Allihn, *Scheibler's Neue Zeit.*, 1879, iii. 230. Soxhlet, *Chem. Centb.*, 1878,

Meissl,* who has used this method of estimating the invert-sugar in raw sugars from the cane, states that in an experiment made by him on a mixture containing known quantities of cane and invert sugar, the former was proved to be without influence upon the results.

B. The Gravimetric Method.

The investigations of Soxhlet (*loc. cit.*) lead him to the conclusion that the gravimetric method for estimating sugar is empirical and quite inexact on account of the inconstancy of the relation between the sugar and the copper oxide reduced. In this particular, however, his views are negatived by the recent exact experiments of Marcker † and Meissl ‡ (*loc. cit.*), together with the uniform experience of many other chemists.

The Estimation.—25 c.c. of Fehling's solution (see for-

Nos. 14, 15; *Stammer's Jahresb.*, 1878, 178; *Jour. Chem. Soc.* [abs.], xxxiv.; *Jour. Pk. Chem.*, [2] 21, 227, 317; *Jour. Chem. Soc.* [abs.], Oct., 1880; Gratama, *Fres. Zeit.*, xvii. 155.

* *Stammer's Jahresb.*, 1879, 180.

† *Chem. Centb.*, No. 37; *Stammer's Jahresb.*, 1878, 189.

‡ Meissl gives only a qualified approval of the gravimetric method, holding that a close and unvarying relation must exist between the quantity of precipitated oxide, the reducing sugar, and the cane-sugar present—such a relation which in practice it would be very troublesome to obtain. He also states that the presence of cane-sugar and the length of time the boiling of the solution is carried on have important influence on the results. He considers the process, as ordinarily carried out, to give figures that are too high. The whole question of the gravimetric estimation of dextrose and invert-sugar is in an unsatisfactory condition, owing to the contradictory nature of the results recently obtained; but the weight of opinion, principally founded on the elaborate investigations of Soxhlet and Meissl, points to the fact that the gravimetric estimation is apt to give high results, especially in the presence of cane-sugar; and in cases where accuracy is of the highest importance it is recommended that the tests be duplicated by this method, and the volumetric according to Soxhlet, giving preference to the latter in doubtful cases.

mula, page 190) are placed in a 100 c.c. flask with the sugar solution of known strength, but not containing more than .120 gramme of the pure sugar, and the volume brought to 100 c.c. The contents of the flask are heated on the water-bath 20 minutes, or until a complete reaction is assured, care being taken at the end of the operation that the copper solution is in excess. Filter off the precipitated suboxide; wash with hot water; dry the filter and burn it separately from the precipitate; ignite both in a platinum boat, or a Rose crucible with perforated cover; and finally reduce by heating to redness in a current of dry hydrogen. After complete reduction, cool as quickly as possible in hydrogen and weigh.

Weight of copper $\times$ .5678 = dextrose or invert-sugar.

Weighing as Oxide.—The precipitated copper may also be determined as oxide by igniting the suboxide and ash, after burning the latter separately, in a platinum crucible with access of air at a strong red heat, cooling, oxidizing with a few drops of nitric acid, again heating for some time, and finally weighing. The cupric oxide multiplied by .4534 gives the invert-sugar or dextrose.

Filter Ash.—The filter generally obstinately retains some of the salts of the alkaline solution, which no ordinary amount of washing seems to be able to remove. Soxhlet has placed this very high—as much as 20 milligrammes for an ordinary-sized filter, though Marcker (*loc. cit.*) has proved it to be very much less. It is advisable to make a special determination of the ash of the filters to be used for this method, on a number through which copper liquor has been run, and the quantity adhering afterward washed out with hot water.

Mohr's Method.*—This is said to give results as accurate as the above with much less expenditure of time. When moist suboxide of copper is mixed with an acid solution of a ferric salt it becomes oxidized to cupric oxide, passing into solution, and reducing an equivalent quantity of the iron compound to protoxide, which may be estimated volumetrically by permanganate of potassium. The suboxide formed by boiling the sugar solution with excess of copper liquor is filtered off, thoroughly washed, and an acid solution of ferric sulphate, or ammonio-ferric alum free from FeO, is poured upon the filter until the suboxide is dissolved. The filter is then washed as long as any iron solution adheres to it, and the filtrate, together with the washings, is titred with a solution of potassium permanganate containing 3.162 grammes of pure salt to the litre.

1 c.c. permanganate solution used = .0036 gramme invert or grape sugar.

There have been proposed several other processes for determining the precipitated suboxide volumetrically with different reagents, and also the modification of adding a known volume of copper liquor to the sugar solution, and after the reaction estimating the amount of copper remaining dissolved after filtering off the suboxide—from which the amount thrown down as suboxide is found by difference. None of these devices, however, have any advantages over those already given in detail.

* Mohr, *Tithirmethode*, 5th Aufl., 449.

SECTION II. DETERMINATION OF DEXTROSE AND INVERT-SUGAR BY OTHER METHODS THAN THAT OF FEHLING.

Knapp's Method.*

This is based on the fact that an alkaline solution of cyanide of mercury is reduced to the metallic state by grape-sugar. It is carried out as follows: 10 grammes of pure and dry cyanide of mercury are dissolved in distilled water, the solution mixed with 100 c.c. of caustic soda-lye sp. gr. 1.145, and sufficient water to make one litre; .400 gramme of the cyanide (= 40 c.c. of the solution) at the boiling-point is completely reduced by .100 gramme grape or invert sugar. The titration is performed as in Fehling's process: 40 c.c. of the mercurial solution is boiled in a porcelain dish, and the sugar solution (not stronger than ½ per cent.) is added as quickly as possible until the metal is all thrown down. It is best to make a preliminary experiment to find the approximate amount of sugar solution required, and then on the second trial nearly the whole amount may be run in at once. To test when the end point is attained, a drop of the liquid is placed on a piece of filter-paper stretched over the mouth of a beaker-glass the bottom of which is covered with a strong solution of sulphide of ammonia. As long as any mercury remains in solution a brownish spot appears on the paper when a drop from the boiling solution is placed upon it. Perhaps a better plan would be to use the alkaline solution of zinc oxide, which may be brought in contact with the solution on a porcelain plate. While mercury yet remains in solution a brownish coloration is produced.

---

* *Ann. der Chem. Pharm.*, May, 1870; *Amer. Chemist*, i. 118.

Knapp, Lenssen,* and Soxhlet, who have carefully examined the method, recommend it.

Sachsse's Method.

A standard solution is made by dissolving 18 grammes of pure, dry mercuric iodide in water with the aid of 25 grammes potassium iodide, adding to the liquid 80 grammes of caustic potash and water to make one litre. 40 c.c. of the solution, containing .72 gramme HgI, is equivalent to .1342 gramme dextrose and to .1072 grm. invert-sugar.

For the estimation, 40 c.c. of the standard solution are boiled, and the sugar solution of known strength added until all of the metal is precipitated. This point is ascertained by bringing a drop of the solution in contact with sulphide of ammonia on a piece of Swedish filter-paper, or by an alkaline solution of zinc oxide, both of which reactions give rise to brownish precipitates or colorations with solutions of mercuric salts.

This process is also based on the reduction of mercury to the metallic state by the sugar present. Heinrich,† Meissl,‡ and Soxhlet § recommend it as giving results closely accordant with those by Fehling's method.

Estimation of Dextrose and Invert-Sugar in the presence of each other.

On account of the peculiarity of Sachsse's method, in that dextrose and invert-sugar have distinctly different reducing constants for the alkaline solution of mercuric

* *Fres. Zeitschrift*, No. 4, 1870.

† *Stammer's Jahresb.*, 1878, 195.

‡ *Ibid.*, 1879, 178.

§ *Journ. Chem. Soc.*, Oct., 1880.

iodide, it is possible, by a combination of it with Fehling's process, to determine dextrose and invert-sugar in the same solution. An example will illustrate the necessary steps of the operation: 25 c.c. of sugar solution was found to reduce 40 c.c. = .72 gramme mercuric iodide, of the Sachsse solution, while the same volume contained, according to the copper test, .125 gramme of the total sugars; then, calling the dextrose x and the invert-sugar y, we get the equation:

$$x + y = .125 \quad (1)$$

But as .1342 gramme dextrose and .1072 gramme invert-sugar each reduce .72 gramme HgI, we have $x \times \frac{.72}{.1342} = 5.36\,x$, and $y \times \frac{.72}{.1072} = 6.70\,y$; whence

$$5.36\,x + 6.70\,y = .72 \quad (2)$$

On combining these equations the values of x and y may be readily obtained. Should x or y be found equal to 0, it indicates that the corresponding sugar is not present.

Cane-sugar in mixture with the other two bodies may also be estimated, where there are no foreign reducing substances present, by determining the dextrose and invert-sugar as above, and then inverting with acids and estimating the amount of invert-sugar formed from the cane-sugar, by the copper method; the result, multiplied by the factor .95, gives the saccharose originally in the solution.

Estimation of Levulose and Dextrose in the presence of each other (a combination of the optical with Fehling's method).—According to Neubauer,* the angular rotatory power in a tube 100 mm. long, at 14° C., of

Dextrose is $[a]\,j = 53.1°$ — Rotation constant, 1883.3.

Levulose, $[a]\,j = 100°$. — Rotation constant, 1000.

* Neubauer, *Ber. Chem. Gesell.*, 1877, 827.

The following table shows the angular rotations of sugar solutions at 14° C. in a 100 mm. tube:

Per cent.		Levulose.	Dextrose.
1	Corresponding rotatory power	− 1°	+ .531°
2	" " "	− 2°	+ 1.062°
3	" " "	− 3°	+ 1.593°
4	" " "	− 4°	+ 2.124°
5	" " "	− 5°	+ 2.655°
6	" " "	− 6°	+ 3.186°
7	" " "	− 7°	+ 3.717°
8	" " "	− 8°	+ 4.248°
9	" " "	− 9°	+ 4.779°

Estimate the total sugars by Fehling's method, and also make an observation with the saccharimeter at 14° C., in a 100 mm. tube, the result being set down in angular degrees. As an example, suppose a solution containing 15 per cent. of total sugars has a rotation of − 5.202°; 15 per cent. levulose corresponds to a rotation of − 15°. The difference between this and the actual rotation is (− 15°) − (− 5.202) = − 9.798°. This difference is caused by the dextrose present, and as the algebraic difference between the rotation constant of levulose and dextrose (2883.3) is to the rotation constant of dextrose (1883.3), so is the difference between the calculated and found rotations (− 9.798°) to the quantity of dextrose present—as,

$$2883 : 1883.3 :: 9.798 : x$$

$$x = 6.4 \text{ per cent. dextrose.}$$

15 per cent. − 6.4 per cent. = 8.6 per cent. levulose.*

* Apjohn (*Chem. News*, xxi. 86) and Dupre (*Chem. News*, xxi. 97) give processes for estimating cane-sugar, dextrose, and levulose in the presence of each other, consisting of combinations of the optical and Fehling's methods.

Gentele's Method.*

An alkaline solution of potassium ferricyanide, when heated with grape or invert sugar, is reduced to ferrocyanide, and the strongly yellow solution is wholly or partially decolorized. A standard solution is prepared by dissolving 109.8 grammes pure potassic ferricyanide and 55 grammes caustic potash in water, and diluting to one litre:

10 c.c. of this solution = .010 cane-sugar;
.01053 grape or invert sugar.

40 c.c. of the standard solution is heated in a porcelain dish to 70° to 80°, and the sugar solution slowly added until the color is discharged.

This process has chiefly an historical interest.

---

* *Dingler's Polyt. Journal*, clii. 68, 130.

CHAPTER IX.

ANALYSIS OF RAW SUGAR.

Composition of Raw Sugars.—Raw sugar as obtained from various sources is a mixture, of which the leading element is cane-sugar; the different constituents may be classified as follows:

1. Cane-sugar.	4. Moisture.
2. Invert-sugar.	5. Organic matters not sugar.
3. Salts.	6. Insoluble constituents.

The *cane-sugar* exists in three forms—viz.: (1) actually crystallized; (2) that capable of crystallizing, but held in solution by the moisture present; and (3) that which is incapable of crystallizing, though chemically identical with crystallizable cane-sugar; it is distinguished from the latter by being soluble in alcohol saturated with cane-sugar. The non-crystallizable cane-sugar, with the invert-sugar and the other soluble impurities, form the molasses, which, with that formed in the process of refining, is part of the yield of the refiner.

The *salts* are organic or inorganic; the former consist of the bases potassa, soda-lime, magnesia, and iron oxide united with the organic acids, acetic, succinic, malic, citric, oxalic, tartaric, aconitic,* aspartic,† pectic, metapectic, lactic, proprionic, butyric, formic, glucic, apoglucic, humic, ulmic, and melassic; the latter, with the same bases, are

* Has been found only in raw cane-sugar and juice. † Peculiar to beet-sugar.

chlorides, sulphates, phosphates, silicates, carbonates, and nitrates. Saccharates of lime and potash are also often present, especially in raw beet-sugars.

The *organic matters not sugar* are: I. The alkaloids—betaine,* asparagine,* and triethylamine; II. Nitrogenous organic matters—albumen, legumine, and ferments or extractive matters; III. Non-nitrogenous organic matters—cellulose, gum and fatty bodies, pectose, pectin, parapectin, essential oils, mannite, alcohol, starch, and coloring matter consisting for the most part of caramel and its derivatives.

All of these bodies are not found in any one sample of raw sugar; many are peculiar to raw beet-sugars, not being found in those derived from the cane.

Insoluble matters. These are principally sand and clay with accidental mechanical impurities; also fine particles of fibre from the cane often occur in raw sugars, giving the solution a muddy appearance.

Sampling.—Much depends upon the proper taking of the sample as well as its preservation. The sampling from the original packages is evidently not in the province of the chemist, and must be left to the experience and intelligence of the sampler, and the business customs which sometimes control this operation. The chemist should, however, when practicable, receive from two to five pounds of each lot to be tested, representing as closely as possible the larger sample first taken; this, if containing lumps, should have them well broken up and the whole carefully mixed. The average sample thus obtained is preserved in a tightly-stopped, wide-mouthed bottle, and from this the various

---

* Peculiar to beet-sugar.

portions should be weighed out for analysis. It is important to have the bottle well corked, as in ordinary states of the atmosphere raw sugar loses its water very rapidly by evaporation, though when the air is damp the reverse may take place, and moisture be absorbed, thus vitiating the analysis in its most important item—*i.e.*, the cane-sugar.

ESTIMATION OF THE CANE-SUGAR.

This determination is made by the optical saccharimeter, and though the choice of any of those described in the preceding pages is open to the operator, for the sake of convenience the description in this section is based upon the supposed use of the Soleil-Ventzke instrument ; this remark will also apply to the matter in relation to the use of polarizing apparatus in other sections of this work.

The Weighing.—The normal quantity, 26.048 grammes, is weighed with the requisite accuracy into a convenient-sized German-silver dish of the form shown in Fig. 27, provided with a projecting lip so that it may be adapted to pouring into the neck of the measuring-flask ; the dish has a counterpoise weight.

Fig. 27.

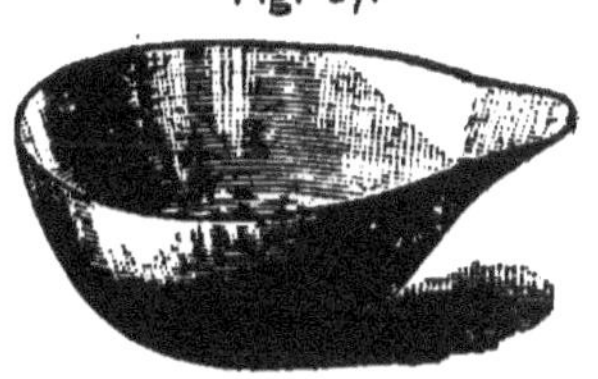

There is also a normal (26.048 grammes) and one-half normal weight. The balance to be used should not be too fine, as much time would be lost in weighing, on account of the care necessary to avoid injury to it, as well as the slowness with which such balances swing. A strong, well-made instrument with a rather short beam, and capable of quick weighing to within .010 gramme of the truth, is in every way suitable; a good apothecaries' balance, accurate to the limit specified, does very well.

The Execution of the Test.—When the assay is weighed, about 50 c.c. of water is poured upon it, and by means of a thick glass rod with a blunt end the mixture is stirred, any lumps present being at the same time crushed, until the greater part of the sugar is dissolved. By allowing the contents of the dish to settle a moment, the solution may be readily poured off into the flask, leaving the undissolved residue of sugar entirely in the dish. A second portion of water is added, and so on until the sugar is completely dissolved, and the dish rinsed out into the graduated flask. There is little advantage to be obtained in the use of hot water, except in the case of crystal-sugars, as the time required to cool the solution to the temperature of the operating-room is about equal to that saved by using heated water. A good arrangement for holding the water to be used in the tests is to have it in a bottle placed on a high shelf, and furnished with a syphon passing through the cork, hanging down the outside within convenient reaching distance of the operator. The syphon is kept filled with water, the lower portion consisting of a rubber tube, the end of which is closed by a spring clamp (Fig. 28). After all the rinsings of the weighing-capsule have been added to the flask, the solution of basic acetate of lead or other clarifying agent is poured in, and the flask filled with water until the lowest portion of the meniscus, or curve formed by the liquid in the neck, is just tangent to the graduation-mark. The solution is then well shaken, a part of its contents thrown away, a little bone-black added, shaken again, and the mixture filtered. Care must be taken that the temperature of the liquid does not differ much from that of the operating-room throughout the manipulations, as an alteration in the volume of the sugar solution is attended with a corresponding change in the amount of sugar as shown by the polarizing apparatus.

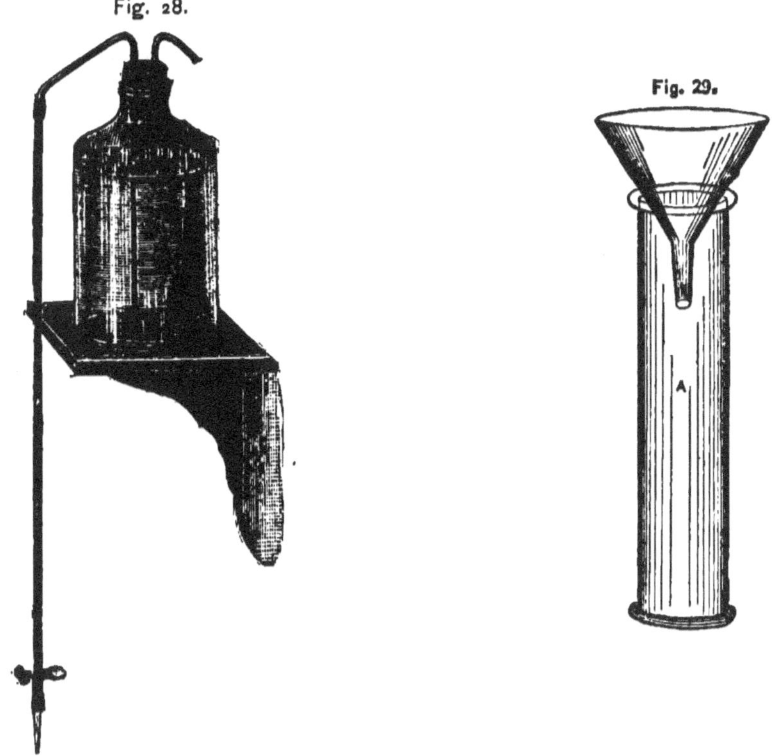

Fig. 28.

Fig. 29.

The quantity of lead solution to be used must be left largely to the judgment of the operator, due regard being paid to the remarks pp. 165, 166. Always employ the least amount with which a good reading in the sacchari-meter may be assured; but not too little, for nothing is gained by attempting to make an observation with a dark solution. When from insufficient decoloriza-tion a filtrate is obtained too dark for reading, a new portion of the sugar should be weighed and the test repeated, or the filtrate from the first solution may be treated with a fresh portion of bone-black and refiltered. The addition of a few drops of alum solution after the lead salt often greatly increases the decolorizing effect.* See page 164.

Any suitable arrangement for filtration may be adopted. A very good one is shown in Fig. 29. A is a cylinder of thick glass 135 mm. high and 27 mm. internal diameter. The funnel is of ordinary glass, 78 mm. in diameter at top, and of a size to fit a filter 150 mm. diameter. It is scarcely necessary to say that filter, funnel, and cylinder should be perfectly clean and dry before use.

When an estimation of the glucose is to be made, by taking out a portion of the solution made up to 100 c.c. for polarizing, as recommended on page 195, the above method of procedure is modified somewhat. Instead of adding the lead to the solution before it is made up to the mark, this is done without any such addition, the solution shaken, and from the lead-free liquid 19.2 c.c., or any other volume equivalent to 5 grammes of the original substance, is taken in a suitably graduated pipette, and diluted for the glucose determination. The remainder of the sugar normal solution is poured into a 50 c.c. flask (after rinsing it out with a portion of the same solution) to the mark, a measured volume of lead solution added from a graduated pipette, and a correction added to the polariscopic reading corresponding to the amount of dilution caused.

It is to be recommended that the long tube of 200 mm.

---

* It has been asserted that the presence of ammonia or ammoniacal salts in the sugar liquid causes an error, owing to the formation of a precipitate of sugar and acetate of lead.

should be used whenever it is possible. For work demanding ordinary care it is generally preferable, in the case of dark solutions obtained by insufficient decolorization, to weigh out a new portion of the sugar, and use more bone-black and lead, than to employ the short or 100 mm. tube with the attendant errors of multiplication. The observation-tubes, as well as the caps and plates, should be scrupulously cleaned after use, by washing and drying with a soft rag or chamois-skin, both inside and outside. Air-bubbles in the flask, which make it difficult to fill the liquid exactly up to the mark, may generally be prevented from making their appearance by adding the solution from the weighing-capsule gently along the neck and side of the flask, and the lead solution in the same manner. The air-bubbles, if present, may be readily removed by allowing a little ether vapor to flow into the measuring-flask from an open bottle, or by the addition of a drop, and blowing off the excess of the vapor.*

ESTIMATION OF THE INVERT-SUGAR.

This determination is made according to the methods given in chapter viii.

ESTIMATION OF THE WATER.

The moisture of raw sugar is determined by drying at such a temperature as will drive off all the water in a rea-

* There has been a method proposed in France for the estimation of cane-sugar in raw sugar, known as the *four-fifths method*, and, I believe, used to some extent in commercial analysis. It consists in taking four-fifths of the ash as the number expressing organic matter not sugar. The sum of this—the ash, water, and glucose—subtracted from 100, represents the cane-sugar. The method is not worth mentioning, except as a curious example of the aberrations to which the human mind is subject.

sonable time without decomposition. The difference in weight of the sugar, plus the containing vessel, before and after desiccation, gives the amount of water. Two or three grammes of sugar are weighed on a tared watch-glass and placed in the drying apparatus. When the desiccation is completed the watch-glass, with contents, are reweighed. Glasses sixty millimetres in diameter are of convenient size. They may be numbered by scratching on them with a hard file, and a memorandum of the corresponding tares kept posted up near the balance.

Example :

Watch-glass + sugar before drying,	10.256	10.256
"	8.120	
Raw sugar..........	2.136	
" + sugar after drying,		10.153
Loss, equivalent to water...........		.103

$$\frac{.103 \times 100}{2.136} = 4.82 \text{ per cent. water.}$$

The sugars should be weighed as quickly as possible, and some varieties, notably large-grained refined sugars and centrifugals, ought to be enclosed between two watch-glasses, as they alter so rapidly by reabsorption after drying that it is sometimes difficult to get a correct weighing in the ordinary way. The heat of the water-bath oven is inadmissible as a means of desiccation when the analyses are made for commercial purposes, as the time necessary for a reliable result is much too great. An accurate estimation in this way can only be made by drying with repeated weighings until the last two closely approximate each other. The time required for this, as my experiments* show, is too long to make the results of the test available commercially.

The drying is best effected in an air-bath at a temperature of 110° C. The time required varies with the kind of sugar operated upon; for dry sugars comparatively free from syrup a much less time will be sufficient than for the moist, low-grade products of the refinery or plantation. No exact rule can be given, but, on the average, good quality of beet and refined sugars, centrifugals or crystal-sugars, good and fair muscovado, good Manillas, Javas, and other similar sugars, will lose their moisture in from two to three hours, while lower grades, such as domestic molasses and low refinery products, Chinese and such sugars, ought to be dried considerably longer, though not too long, for it is an observed fact that impure sugars are altered in composition, attended with loss in weight, by very long continued heating, even at a temperature as low as 95°.

For very syrupy sugars and melados it becomes necessary to dry with the addition of sand. The operation is con-

---

* The leading authorities attach too little importance to the fact that the water in raw sugar exists as a component of a dense syrup, and consequently the last portions of water resist the desiccation with great obstinacy. As an average of many hundreds of dryings at 95° C. (the temperature of the water-bath oven), I find that the following times are necessary for the complete desiccation of such sugars (cane) as occur in the markets of this country:

(Quantity dried of each kind, about two grammes.)

Centrifugals.................	9	hours.	
Ordinary muscovados........	8½	"	
Low "	8	"	
Domestic molasses sugar, made fr. W.I. molasses....	40	"	(alteration).
Manillas.....................	8 to 4	"	
Melado in sand..............	16	"	"
A refined....................	7	"	
C " (lowest product)..	5 to 6	"	
D " (lowest product)..	40	"	"

ducted as follows: Weigh in a good-sized watch-glass or metal dish a small glass rod and about ten grammes of clean, coarse sand which has been ignited and preserved in a tightly-closed bottle; after the combined tare is taken, add about two grammes of the substance to be examined, and, after weighing again, carefully mix with the rod,

Fig. 30.

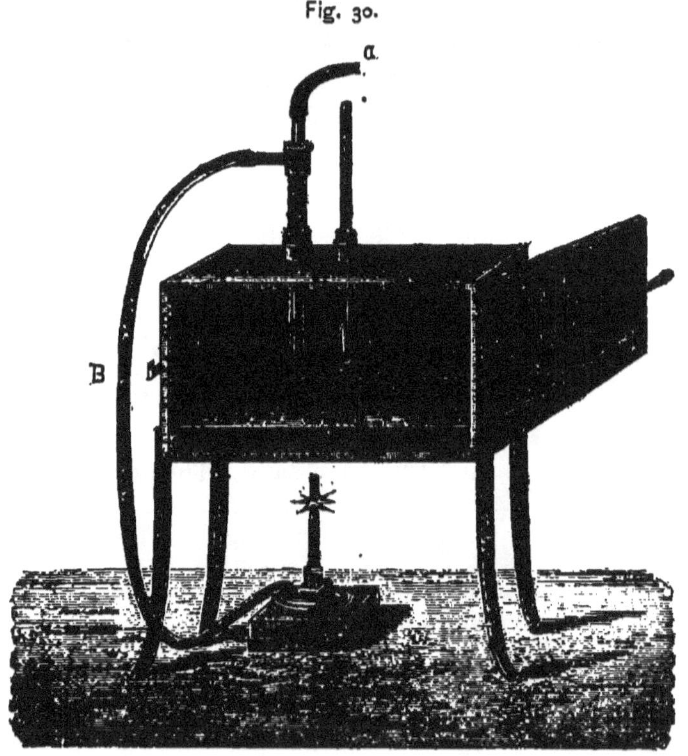

moistening with a few drops of alcohol, so that the assay is thoroughly and evenly mixed with the sand and none of the latter is lost; the rod, after the stirring, is allowed to remain in the dish and is weighed with it.

The best method of drying syrupy sugars, whether with sand or not, is in a vacuum at 90° C., as not only is the operation shortened, but the alteration of the sugar by heat is reduced to a minimum. Scheibler has proposed an apparatus for drying *in vacuo.** Sugars that are nearly free from invert-sugar, such as high-grade refined and crystal raw sugars, may be safely dried in the ordinary way at 120°; and, indeed, that temperature is often necessary to complete the drying in a reasonable time.

Fig. 30 shows the air-bath furnished with a Bunsen regulator, whereby the flow of gas may be made so that any desired temperature may be kept constant; the gas enters at *a*. If the temperature is too high the mercury in the bulb of the regulator rises and partially cuts off the flow of gas; on the contrary, when the temperature in the bath becomes lowered, the mercury recedes and the supply of gas is increased. In this way the heat may be kept pretty equable as long as the pressure of the gas remains the same. The samples to be dried are placed on a perforated shelf raised some distance above the bottom, and the bulb of the thermometer should nearly touch the former. Gas-regulators are, however, often unsatisfactory in practice; generally no difficulty will be encountered in maintaining a constant temperature, without a regulator, with an air-bath modified from the one shown in Fig. 30 by being enclosed below with a sheet-copper box having a door for the purpose of lighting the Bunsen burner; the whole arrangement may be screwed against the wall in a place sheltered as much as possible from draughts. With a little experience in regulating the size of the flame, this apparatus gives excellent results.

The Balance.—The weighing apparatus suitable for the

---

* *Stammer's Jahresb.*, 1870, 199.

estimation of water, ash, etc., should be accurate to .0005 gramme, and not too slow in movement.

ESTIMATION OF THE ASH.

The ash of raw sugar represents its fixed mineral constituents, and is a part of the salts present; these salts are combinations of organic and inorganic acids, and radicals, with various bases. When the sugar is incinerated, the organic matter, including the sugar, is oxidized, and a part of the carbonic acid formed unites with the alkaline and earthy bases, producing carbonates, which, together with sulphates, chlorides, silicates, etc., constitute the ash. The different combinations, and their proportions relative to each other, vary much according to the conditions—viz.: 1. The source of the sugar, whether from the cane, the beet, or of any other origin. 2. The character of the soil, climatic conditions, manures, and the process of manufacture. This last item will of itself furnish a wide range of variation, as chemicals are in use at the place of manufacture which in many cases remain in the raw products, and necessarily modify their saline content. What might be called the *normal ash* of raw sugars, is that from sugars in which no chemical has been used in the course of manufacture, except lime.

The ashes of beet and cane sugars differ materially; in the former there is a large preponderance of the salts of potassium and sodium, while the latter are characterized by a much smaller quantity of alkaline, but more lime-salts and silica. The insoluble impurities, such as clay, sand, etc., are more common in cane than in beet sugars. The following analyses give the composition of raw cane and beet sugar ash:

BEET-SUGAR ASH. Average Composition.

I.

Carbonates of potassium and sodium	82.20
Calcium carbonate	6.70
Potassium and sodium sulphates and sodium chloride.	11.10
	100.00

—(Monier.)

II.

	By simple incineration.		By addition of SO_4H_2.
Sulphuric anhydride	17.63		58.38
Chlorine	4.48		*....
Silica	0.72		.72
Carbonic anhydride	22.87		*....
Lime	6.53		6.53
Potash	25.65		25.65
Soda	21.62		21.62
	99.50		112.90
Undetermined and loss	.50	less $\frac{1}{10}$	11.29
	100.00		101.61

—(Scheibler.†)

III. Mixture of ash from a refinery laboratory accumulated in one year's work, and hence probably a very good average.

Potash	34.19
Soda	11.12
Lime	3.60
Magnesia	.16
Ferric oxide and alumina	.28
Sulphuric anhydride	48.85
Sand and silica	1.78
	99.98

—(J. W. McDonald.‡)

RAW CANE-SUGAR ASH. Average Composition.

I.

Carbonate of lime	49.00
Carbonate of potassium	16.50
Sulphates of sodium and potassium	16.00
Chloride of sodium	9.00
Silica and alumina	9.50
	100.00

—(Monier.)

* Driven off by sulphuric acid.

† *Stammer's Jahresbericht*, iv. 225.

‡ *Chem. News*, xxxvii. 127.

II. Demerara sugar containing 1.38 per cent. ash of the following composition:

Potash	29.10
Soda	1.94
Lime	15.10
Magnesia	3.76
Sulphuric anhydride	23.75
Phosphoric anhydride	5.59
Carbonic acid	4.06
Chlorine	4.15
Ferric oxide	.55
Alumina	.65
Silica	12.38
	101.03
Deduct oxygen equivalent to chlorine	.93
	100.10

The juice from which this sugar was made is supposed to have been treated only with lime (Wallace).*

III. Sulphated ash from one year's work in a refinery laboratory:

Potash	28.79
Soda	.87
Lime	8.83
Magnesia	2.73
Ferric oxide and alumina	6.90
Sulphuric anhydride	43.65
Sand and silica	8.29
	100.06

—(J. W. McDonald.†)

The estimation of the ash is made by incinerating from two to three grammes of the sugar in a small platinum dish at a red heat; the difference in weight of the dish before and after ignition gives the absolute quantity of ash, which multiplied by 100 and divided by the amount of the assay gives the percentage.

Soluble Ash.—A simple incineration of the sugar gives the total ash, regardless of its composition. As insoluble

---

* *Chem. News*, xxxvii. 76. † *Ibid.*, xxxvii. 127.

matters, like sand and clay, do not exert any injurious action on the sugar in the process of refining, it is often desirable to know the soluble part of the ash. This may be determined as follows: Weigh from two to five grammes of substance, dissolve in a little boiling water, filter hot, wash with hot water, and evaporate the filtrate and washings in a tared platinum dish; ignite the dry residue, burn off the carbon, and weigh.

Alkaline Ash.—On account of the great volatility of the alkaline carbonates which form a large portion of sugar ashes, the heating for a sufficient time to oxidize the carbon will result in a considerable loss by volatilization, rendering the result of the estimation too low. It is well known that in general the salts of the alkali metals have a powerful melassigenic effect in preventing cane-sugar from crystallizing, and in inversion, while many of the other constituents of sugar-ash are almost inert in this respect (see page 64). To be able to estimate the amount of alkaline salts in a given ash is, therefore, a *desideratum*. This result may be reached in the following way: Thoroughly carbonize the sugar; transfer the coal to a small mortar, pulverize; wash well with hot distilled water and filter; evaporate the filtrate and washings, and dry the residue on a water-bath; or determine the amount of alkaline carbonates by standard acid solution in the ordinary process for alkalimetric estimation (page 258). By the carbonization, the salts are mostly transformed into carbonates, and during the lixiviation the comparatively non-injurious lime salts remain on the filter in great part, together with sand, clay, alumina, magnesium, carbonate, and other matters. Small quantities of caustic lime and other bodies are dissolved along with the alkaline carbonates, owing to the reduction of sulphates to sulphides, but by far the greater portion of the dissolved salts consist of sodium and potassium carbonates, and alkaline chlorides.

Sulphated Ash.—Scheibler * has proposed the incineration of sugars with the addition of concentrated sulphuric acid. The advantages of this method are (1) that the carbonization in the presence of the acid furnishes a porous coal which burns off rapidly and without much swelling, while in the ordinary way the charred mass is apt to become hard and graphitic; (2) that the bases are converted into sulphates, with the expulsion of chlorine and carbonic acid, whereby the loss by volatilization is greatly diminished, as the sulphates are very stable at a red heat compared with chlorides or carbonates. The ash is thus determined: Two to three grammes of sugar are weighed in a small tared platinum dish, as shown in Fig. 31, 45 mm. in diameter, 14 mm. high, with a flat or convex bottom; from fifteen to thirty drops of pure concentrated sulphuric acid are added to the sugar in the dish, and the heat applied at first rather gently, and then to full redness. Nothing is gained by having too large a flame, as the carbon burns off most rapidly with a moderate red heat. Scheibler has proposed the use of a platinum muffle, which, when many determinations are to be made, will be found useful. It is figured at 32. The dimensions of the muffle to hold three dishes, as described above, are 150 mm. length, 55 mm. width, and 25 to 30 mm. high. It should slightly taper towards one end, which is elevated somewhat so as to allow the air a good draught through the apparatus. When

Fig. 31.

* *Stammer's Jahresb.*, iv. 221; vii. 267.

the carbon is completely burned off, the dish is allowed to cool and reweighed. If the amount of ash is large, or it is very fusible, it will often happen that the last portions of carbon are oxidized with difficulty. In such a case the dish is allowed to cool, one or two drops of sulphuric acid added, and the dish heated cautiously at first to avoid spattering, and finally brought to redness for fifteen minutes.

As the equivalent of sulphuric acid is greater than

Fig. 32.

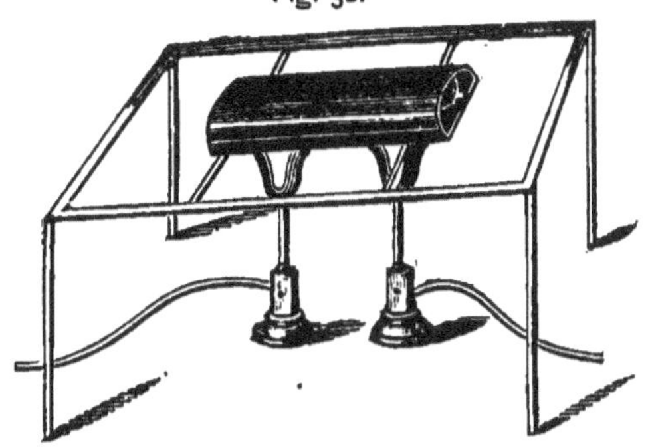

that of carbonic acid (in proportion of 40 : 22), it is necessary to reduce the net weight of the sulphated ash to the figure that represents the carbonated ash. Scheibler has found that a subtraction of one-tenth from the weight of the sulphated ash will do this approximately. Though the discrepancy mentioned above is not constant for all sugars, yet the results given by the method are near enough to the truth for all practical purposes.* Example:

Sugar plus dish	12.121
Dish	10.110
Sugar taken	2.011

* Violette (*Amer. Chemist*, v. 296) considers that the coefficient $\frac{2}{10}$ should be used in general, and $\frac{3}{10}$ for very pure raw sugars.

Dish plus ash..	10.120		
Dish...........	10.110		
Ash............	.010	$\frac{.009 \times 100}{2.011}$	= .44 per cent.
Less $\frac{1}{10}$.........	.001		
	.009		

The process with sulphuric acid is preferable to the simple incineration for accuracy, general agreement of results, and facility of execution. The *soluble ash* can also be made by the sulphuric-acid method. It is to be recommended that with sugars containing much sand, clay, and other insoluble impurities, the soluble ash be taken, as this represents the amount of the salts which go into solution in the operation of refining.

SCHEME FOR THE EXAMINATION OF SUGAR-ASH.

Dissolve 10 grammes of the sugar in water, dilute to 100 c.c., and filter:

A. Dry filter; ignite in tared platinum dish; subtract filter ash. Residue = **Sand, Clay, etc.**	B. Evaporate 25 c.c. (= 2.5 grammes sugar) of the filtrate in a tared dish, add sulphuric acid, carbonize, burn off coal, and weigh; subtract $\frac{1}{10}$ from residue. **= Soluble Ash.**	C. 50 c.c. (= 5 grms. sugar) of filtrate are evaporated in a platinum dish, carbonized, the coal washed with hot distilled water, and the washings, after filtration, evaporated on water-bath to dryness. Result = **Alkaline Ash.*** D. The alkaline ash is titred with standard acid. Result = **Alkaline Carbonates.**

* Alkaline chlorides and carbonates.

The relation between the sulphated ash and the salts as they exist in the raw sugar is such, according to Landolt, that one part of the former is equivalent to two parts of the latter. This is for beet-sugars.

ESTIMATION OF THE COLOR.

Several instruments have been invented for effecting a comparison of color between sugar solutions. Of these Payen's decolorimeter and its modification, Ventzke's, have for their object the estimating of the decolorizing power of char. Salleron's, Duboscq's, and Stammer's colorimeters, and Stammer's chromoscope, are more general in their application, and permit the color-comparison of all saccharine products, solid and liquid, as well as the estimation of the decolorizing power of animal black. None of these appliances, with the exception of Stammer's colorimeter (*Farbenmäss*), have a standard of comparison in the instrument itself; the results are merely comparisons with standard solutions of caramel, which in practice are found to be exceedingly difficult to make twice alike by the same operator. Hence estimations made with such standards have necessarily a considerable element of uncertainty in them.

Stammer's Colorimeter (not to be confounded with his chromoscope) approaches nearest to an absolute standard, the results obtained by different instruments and operators by Stammer's process being generally strictly comparable. The apparatus is shown in Fig. 33. The solution-tube I is closed at its lower extremity by a glass plate, and is open above, where it is provided with a lip by which the sugar solutions may be poured. I is fixed to the wooden support by two screws, which, when it is necessary to clean the instrument, can be easily removed. The measuring-tube III is closed below by a glass plate, and moves freely up and down in the solution-tube I. II, fastened to III, is open below, and at its upper extremity is covered by the colored glasses which form the standard of comparison; at the lower portion of it are two rings with screws, which are connected with a slide moving in a groove cut in the wooden support shown in the figure. The slide serves as an indicator of a millimetre scale placed at the back of the wooden frame, for the purpose of measuring the perpendicular distance that the joined tubes, II and III, may be raised.

Fig. 33.

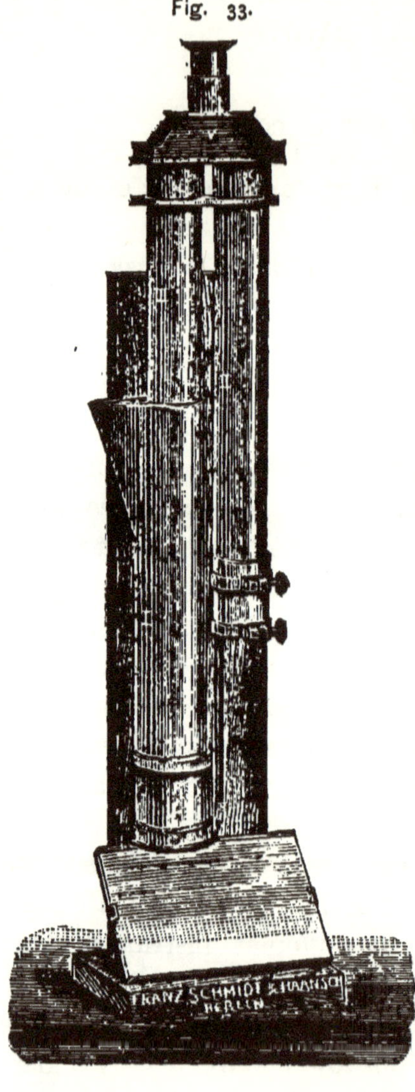

The standard consists of two glasses, and a degree of color equal to them is called 100. Besides this, the apparatus is provided with two separate colored glasses, each equal to one of the plates forming the standard, and may be employed in the place of the standard, being equivalent to one-half of it in color intensity; also one or both may be used with the standard glasses, when the combinations are equal respectively to one and a half times, and double the normal standard. The eye-piece V consists of an optical arrangement whereby the color due to the solution under examination, and that from the colored glasses, are made to appear on either side of a vertical line dividing a circular disk, making a luminous field similar to that of the Soleil saccharimeter. In this manner an accurate comparison of color may be obtained. The eye-piece can be fitted on the top of the tubes II and III after the standard glasses are placed in position in the upper part of II. A mirror at the bottom reflects the light upward through the tubes, and a screw behind the wooden frame attached to the tube II enables the operator to elevate or depress at will II and III. The whole apparatus is mounted on a wooden stand, as shown in the cut.

The manner of using is as follows: The operator fills the solution-tube to the proper height with the liquid to be examined, and then, looking through the ocular, by means of the large screw attached to the frame elevates gradually the tubes II and III, until after repeated trials the two halves of the luminous disk appear of the same intensity of color. At this point the screw is turned so as it keeps the apparatus in the position thus obtained, and the reading of the scale taken, which shows the amount of perpendicular elevation. The more III has been raised, the greater the depth of the column of liquid between the bottoms of I and III. The color-intensity of this column is compared with standard glasses. A solution before use must be rendered perfectly clear, by filtration if necessary.

The color of a solution is in inverse ratio to the length of a column of it necessary to produce a given color. If the comparative color be expressed by 100, it follows that the readings in millimetres must be divided into 100 to get the figure expressing the relative color.

The apparatus may be cleaned by loosening the screws holding the rings on the bottom of II, when the latter can be raised out of the solution together with the tube III.

Calculation.—The estimation of the color of raw sugar, *Füllmass*, or other material can be calculated on one hundred parts of cane-sugar contained, and the result shows the relation of color to the saccharimetric strength. A solution of the substance to be estimated is made by dissolving a known weight in water and making the solution up to 100 c.c. It is convenient to take the normal solution intended for the polariscope. The clear solution is placed in the colorimeter and the reading taken, which is divided into 100. If the solution is too dark for use with the standard, one or both of the extra colored plates may be put in and the readings (before division into 100) divided by 1½ or 2; cr the dark solution may be diluted to twice, four times, or any desired volume, the reading being divided by 2, 4, etc., to reduce to the standard of the colorimeter. If, on the other hand, the solution is too light, the standard glass may be replaced by one of the extra glasses and the reading multiplied by two. These directions apply to the use of the colorimeter, whether for solids or liquids. *Example:* 15 grammes of a raw sugar polarizing 85 were dissolved in water and the solution filtered, after making up to 100 c.c. On trial with the colorimeter it was found to be too dark, and the two extra glasses were put in, when the reading was 36. The calculation would then be as follows: $\frac{36}{2} = 18 = \frac{100}{18} = 5.55$, which is the number expressing the color corresponding to $\frac{15 \times 85}{100} = 12.75$ grammes cane-sugar in 100 c.c. of solution. Now, as $12.75 : 5.55 = 100 : x$, $x = 43.5$; which is the color corresponding to one hundred parts of cane-sugar.

Monier's* Method with Standard Colors.—For those not having a colorimeter this process may be found useful, though in every way less satisfactory than the preceding. A series of ten standard colors are prepared by dissolving known weights of caramel in a fixed volume of water, say 25 c.c., in arithmetical progression, the first tube containing one part caramel, the second two parts, the third three parts, and so on. In order to make a comparison of raw sugar by this method five grammes are weighed, dissolved in water, and the solution made up to the bulk of the standard solutions, after filtration. A comparison is now made between the raw sugar solution and the standards, the one it most nearly approaches in color being that which contains the same amount of coloring matter as the raw sugar. For preparation of the caramel see page 331.

ESTIMATION OF THE ORGANIC MATTER NOT SUGAR.

In commercial analysis these bodies are determined by difference, the sum of the sugar, grape-sugar, water, and ash being subtracted from one hundred, and the remainder called the *organic or undetermined matters.* Included in the above term is a great variety of substances, nitrogenous and non-nitrogenous, of which the chief are organic

---

* *Guide pour l'essai et l'analyse des sucres.* Paris.

acids combined with bases found in the ash, gum, coloring matter, albuminous bodies, and insoluble organic matters, as particles of cane or beet, and cellulose. Some of these bodies are inert in their action on cane-sugar in the process of the manufacture or refining of sugar, while others are very injurious, such as the gummy matters, in hindering or preventing crystallization, and the protein compounds, which tend to set up fermentation of various orders in sugar liquids. Though for most commercial purposes the estimation by difference is sufficient when all the other determinations are made correctly, yet in some cases it is desirable to estimate directly the organic substances, and to discriminate, if possible, between them in regard to their greater or less injurious action on sugar solutions. The method by difference is open to the objection that all the errors of the other determinations fall upon the undetermined matters and make it too high or too low, as the case may be. This fact greatly lowers the value of the figures representing the organic substances in many commercial analyses.

Walkoff's Method.—This is based on the fact that tannin precipitates from raw sugar solutions most of the nitrogenous matters and some other bodies. Two grammes of pure dry tannin are dissolved in distilled water, and the volume made up to one litre; 1 c.c. of this solution contains .002 gramme tannin. About five grammes of the sugar to be tested are dissolved in 200 c.c. of water, the solution heated moderately, and the tannin added from a burette. A flocculent precipitate forms, which gradually settles. From time to time a small portion of the liquid is taken out, filtered after the manner described under Estimation of Grape-Sugar, page 197 in connection with Fehling's solution, and a drop of a solution of ferrous sulphate is added to the filtrate. As soon as a dark color is produced in contact with the iron salt the tannin is in excess, and the end point of the reaction is attained. The weight of tannin employed, calculated from the number of cubic centimetres used, divided by six, represenst the amount of organic matters precipitated. The sugar solution should be perfectly neutral. The relation between the tannin and the organic matters precipitated by it, given above, was obtained for beet products, and it is probable that for those of the cane the proportion is different. When the process is used for the latter, the relation might be determined by precipitating an impure sugar solution with a known quantity of tannin insufficient to completely throw down the matters in solution, collecting on a weighed filter, drying at 100°, and calculating the amount of tannin corresponding to the other substances.

Walkoff's process, though somewhat empirical, is capable of giving good *comparative* results.*

Subacetate of Lead Method.—The basic acetate of lead, it is well known, precipitates a large portion of the organic substances present in raw sugars. Besides the nitrogenous bodies precipitable by tannin, gummy and coloring matters, and many organic acids, are carried down. Twenty grammes or more of the sugar are dissolved in a moderate quantity of warm water, and an excess of solution of lead subacetate added; after heating a few minutes the solution is filtered, the precipitate thoroughly washed, diffused in water together with any portions of

* Pellet and Pelton, as the result of an exhaustive examination of the action of tannic acid on beet-molasses, consider Walkoff's process unreliable, as asparagine is not precipitated.

the filter from which it is difficult to detach the precipitate, and treated with gaseous sulphuretted hydrogen until the lead is all thrown down as sulphide, leaving the organic substances that were combined with the oxide of lead in solution. The precipitated plumbic sulphide is filtered from the solution, washed, and the combined washings and filtrate evaporated to dryness in a tared dish on a water-bath, the heating being continued till the mass ceases to lose weight. This method, though somewhat tedious to execute, may furnish results of comparative value.

Schrötter and Monier have proposed a volumetric method with permanganate of potassium, but it is of very doubtful advantage. For the separate estimation of the organic matters in raw sugar products see Laugier, *Guide pour l'analyse des matières sucrées.**

ESTIMATION OF INSOLUBLE MATTER.

These substances include particles of cane or beet fibre, accidental organic or inorganic impurity, sand, clay, etc. 20 to 50 grammes of the sugar are taken, dissolved in boiling water to make a rather dilute solution, which is filtered through a tared filter, by the aid of a vacuum-pump if necessary. After washing sufficiently, the filter is dried at 100° C. until it ceases to lose weight, and the final weight, after the subtraction of that of the filter, gives the amount of *insoluble impurities.* To find the proportion of the organic and inorganic constituents, the filter is burned to an ash in a weighed crucible; after the subtraction of the weight of crucible and filter ash, the remainder is the *inor-*

* *Zeit. f. Rubenz*, xxviii. 805; *Stammer's Jahresb.*, xviii. 222; Bittman, *Stammer's Jahresb.*, xix. 240.

ganic insoluble impurities (sand, clay, etc.) The difference between the latter and the total constitutes the *organic insoluble impurities.*

ESTIMATION OF THE YIELD.

It is important to be able to estimate the amount of cane-sugar obtainable in refining from a given sample of raw sugar. The buyer or seller who has no knowledge of chemistry finds it very convenient to make use of a single figure summing up the results of the chemical analysis which, perhaps, he is able to but imperfectly interpret. It has long ago been observed that two raw sugars having the same polarization give quite different results in refining, as to yield in crystallizable sugar; and this is rightly attributed to the varying nature and quantity of the impurities, which either tend to destroy the cane-sugar by inversion or to prevent its crystallization.

Method of Coefficients.—It is also a well-recognized fact that saline matters have a particularly injurious effect on the cane-sugar in refining, and that in the syrups from which no longer any sugar can be crystallized, there is a more or less fixed relation between the salts and the uncrystallizable cane-sugar. These considerations gave rise to the method of valuing raw sugars that is in extensive use in France, and, somewhat modified, is adopted in the French government laboratories for sugar analysis. This method assumes that for every part of ash in the raw sugar five parts of cane-sugar are prevented from crystallizing, and that for each part of glucose or grape-sugar one or two parts (according to commercial convention) are carried permanently into the molasses. Thus, a sugar containing

Sugar................	92.00
Glucose.............	2.00
Ash..................	1.00

by the method of the coefficients would give

$$(1 \times 5) + (2.00 \times 1) = 7, \text{ or } (1 \times 5) + (2.00 \times 2) = 9,$$

which, subtracted from the amount of cane-sugar, shows a yield of 85 or 83. The above is the method in the form most used, though many have considered the coefficient 5 too high, and figures varying from 3.5 to 5 have been proposed, and to a certain extent adopted. In raw beet-sugars containing very small quantities of grape-sugar the glucose factor is neglected.

A commission appointed by the French government, composed of MM. Aimé Girard, De Luynes, and other chemists, have recommended this mode of valuing raw sugars, which has been adopted, and is now the officially recognized method. The scheme submitted by the above chemists is as follows: From the percentage of cane-sugar given by the saccharimeter is subtracted the sum of—

(1) Four times the weight of the ash (ash burned with addition of sulphuric acid, and one-fifth subtracted);

(2) Twice the glucose when the titre is 1 per cent. or above; the glucose multiplied by 1 when the titre is between 1 per cent. and ½ per cent.; when the titre is below ½ per cent. the glucose is neglected;

(3) 1½ per cent. for waste in refining.

Thus the sugar whose analysis was given above would show a yield of

$$(1 \times 4) + (2.00 \times 2) + 1.50 = 9.50;\ 92.00 - 9.5 = 82.5.$$

The method of coefficients described, and used in France

for the commercial valuation of raw sugars, though doubtless justified for certain beet and high-grade cane sugars, is open to serious objection. The results given by it necessarily vary a great deal, approaching near the truth for some, but falling far short for others, being generally too low. On this account the system has never obtained outside of France. The various saline impurities have individually very unequal injurious effect on cane-sugar, some being almost inert and others very hurtful; besides which the organic impurities have also an injurious action. The soluble portion of the ash, the only one that can have any melassigenic action, in raw cane-sugars is frequently not more than one-half to three-fourths of the total, while with raw beet-sugars nearly the whole is soluble, and consists largely of the most melassigenic salts—namely, those of potassium. Further, the ash in all raw sugars varies with many circumstances—the methods of manufacture, the soil, manure, etc.—and to lay down a hard-and-fast rule to measure its injurious action is not only empirical, but, from the nature of the case, must be very unreliable. The error of the method in giving results that are too low is much more apparent with raw cane, than with raw beet sugars. Take, for example, a number of type analyses of Cuba sugars:

	Good centrifugal.	Fair centrifugal.	Good muscovado.	Fair muscovado.	Molasses-sugar.
Sugar	96.50	92.50	90.00	86.00	81.50
Glucose	.80	2.10	3.00	4.50	6.50
Organic matter, etc.	.60	1.65	1.25	2.00	4.00
Ash	.35	.55	.75	1.00	1.50
Water	1.75	3.20	5.00	6.50	6.50
	100.00	100.00	100.00	100.00	100.00
	(.35 × 4) + (.80 × 1) + 1.50 = 3.70 96.50 — 3.70 = 92.80, yield.	(.55 × 4) + (2.10 × 2) + 1.50 = 7.9 92.50 — 7.90 = 84.60, yield.	(.75 × 4) + (3.00 × 2) + 1.50 = 10.50 90 — 10.50 = 79.50, yield.	(1 × 4) + (4.50 × 2) + 1.50 = 14.50 86.00 — 14.50 = 71.50, yield.	(1.50 × 4) + (6.50 × 2) + 1.50 = 20.5 81.50 — 20.50 = 61.00, yield.

It will be readily seen by those who are experienced in these matters that the above yields for the higher grades, more closely approximate the actual refining results, while for the lower grades the calculated yield falls much short of the actual.

PAYEN'S PROCESS, MODIFIED BY SCHEIBLER.

(*Raffinationwerths—Rendements.*)

Raw sugar, washed with alcohol of about eighty-five per cent. saturated with cane sugar, is deprived of its syrup. This consists in part of glucose, and partly of cane-sugar that has lost the ability to crystallize owing to the presence of various foreign bodies, together with most of the other impurities, as coloring matters, salts, etc. The residue from this operation is the *cane-sugar actually crystallized in the raw product, plus the cane-sugar held dissolved in the water present, and precipitable by the washing solutions.*

Payen's original method is executed in the following way: The wash-liquor is made by saturating one litre of 88 per cent. alcohol to which 50 c.c. of strong acetic acid is added, with finely-pulverized cane-sugar. The object of the acid is to decompose sucrates and render the saline matters more soluble. Ten grammes of the sugar to be examined is first treated with absolute alcohol to deprive it of water, and then with 50 c.c. of the alcoholic sugar solution, the assay being placed in a small tube the solutions poured upon it and allowed to filter through. A second and third washing is given if necessary, the last wash-liquor consisting of 96 per cent. alcohol saturated with sugar. The purified material is then brought on a tared filter, dried at 100°, and weighed.

The method remained in this form for many years and was but little used. In 1871 Dr. Scheibler, of Berlin, in competition for a prize offered by the Society of the Beet-Sugar Industry of the Zollverein, revived the process of Payen, and so improved the manner of executing it as to make it a practically useful method.

It cannot be properly claimed that Payen's method gives the absolute yield that raw sugars will show in refining, for that depends not only upon the manner of working, whereby greater or less perfection is attained, but also upon the fact that the organic or inorganic impurities may differ in amount or kind independently of the percentage of crystallizable sugar present. For example, two raw sugars giving a yield by this process of 90, the one containing one per cent. of total impurities of a slightly melassigenic nature, and the other three per cent. of a more injurious character: it is evident that the amount of sugar obtainable in refining will be very different from the two sugars. As a relative standard, however, the method, when properly executed, is capable of giving valuable information in regard to the worth of raw sugars, as, *cæteris paribus*, the more crystallizable sugar present the greater the yield. The only legitimate interpretation of the results by Payen's method is to consider the latter as an analytical process for determining the *total cane-sugar present, exclusive of that permanently in the form of syrup.*

The following is a description of the manner of making the estimation with the improvements of Scheibler and others.* There are four liquids used, viz.:

No. 1, consisting of 85 per cent. alcohol to which 50 c.c. of acetic acid is added to the litre, and the mixture allowed to stand in contact with a large excess of powdered sugar for a day, being shaken at intervals.

No. 2. Alcohol of 92 per cent. saturated, as above, with sugar.

No. 3. Alcohol of 96 per cent., also saturated with sugar.

No. 4. A mixture of two-thirds absolute alcohol and one-third ether.

The solutions 1, 2, and 3, after saturation, are preserved

---

* *Stammer's Jahresb.*: Scheibler, xii. 179, 195, 211; xiii. 144, 148; xiv. 139; Kohlrausch, xii. 203; xiii. 152; Bodenbender, xii. 196, 207; Lotman, xiv. 145. *American Chemist*, iii. 330; iv. 85.

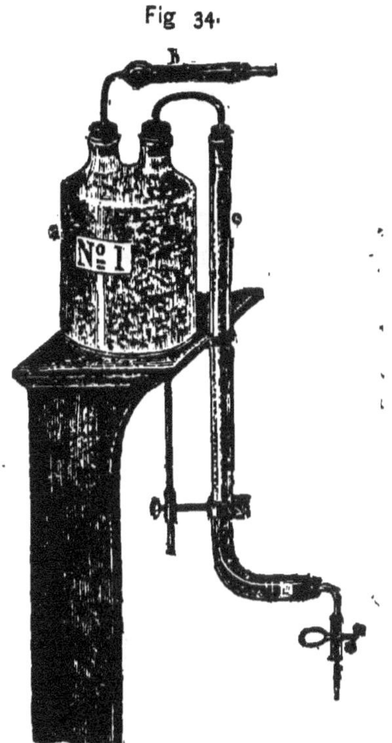

in the two-necked bottles shown in Fig. 34, provided with a syphon delivery-tube, *c*, for drawing off the solution. The bottles are loosely filled with lumps of pure white sugar, as is also the syphon ; *b* is a chloride of calcium tube to prevent moist air from entering. The solutions may be saturated with sugar by allowing them to stand in contact with a large excess of the pulverized substance, and agitating at intervals until the operation is complete. The bottle for holding No. 4 is similar to the above, the syphon being of much smaller calibre. The washing-liquids should be placed conveniently for use in a situation as little liable to changes of temperature as possible.

The washing of the raw sugar takes place in the apparatus figured at 35, one-fourth of the natural size. A is a flask graduated to 50 c.c., which is closed by a rubber stopper of two perforations, one carrying the tube *n*, through which the wash-liquids are added, and another, *o*, which reaches nearly to the bottom of the vessel, and is enlarged at its lower extremity, as shown in the cut. This enlargement serves to hold the filtering material, which consists of little cylinders of the felt used by pianoforte manufacturers, and which fits tightly in the tubes. B is a flask which acts as a reservoir for the solutions after they have been in contact with the raw sugar in A, and from which they are drawn off through a rubber tube connecting with the flasks, by a suction applied to B by a small tube as shown. An ordinary Bunsen water-air pump, or any other arrangement capable of providing a moderate exhaust, is suitable for the purpose.

Fig. 35.

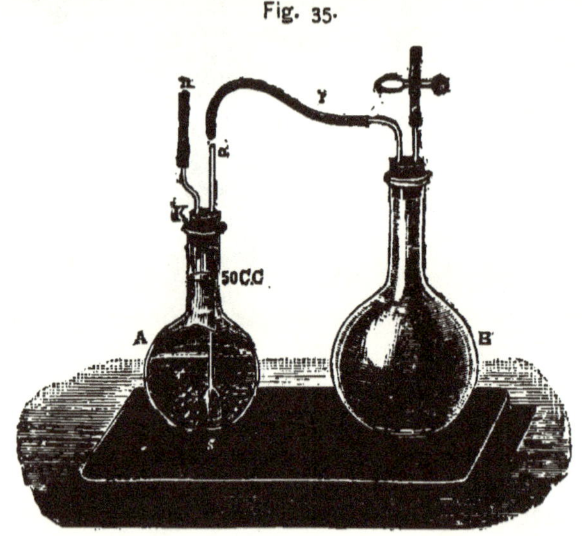

Execution of the Test.—The sample of raw sugar to be tested should be ground in a mortar, if necessary, to break up all small lumps. The half normal quantity of the Ventzke-Scheibler saccharimeter * is weighed (13.024 grammes) and transferred to A. In the case of very moist sugars that would stick to the weighing-dish, it would be better to weigh directly into A, previously tared. The first step of the washing is to run from No. 4 a volume of liquid equal to twice that of the sugar, and allow it to stand

* Or any other quantity to suit the saccharimeter used.

for ten minutes, with frequent agitation to thoroughly disintegrate the mass of sugar and to allow the alcoholic mixture to do its work well. The object of this operation is to remove the water and at the same time to precipitate any cane-sugar dissolved in it. If the acid solution (No. 1) were allowed to act directly on the moist sugar, it would be so diluted by the water present as to be capable of dissolving cane-sugar, and hence make the result too low. If the sugar contains over four per cent. moisture, it is advisable to partially dry the samples after weighing. When the alcohol and ether have remained long enough, the tube *r* is connected with *o*, and by means of the air-pump the liquid is drawn into B; then solutions 2 and 3 are added successively to A, shaken up with the sugar, and similarly withdrawn. The object of the last two solutions is to take up the last traces of alcohol and ether. Solution No. 1 is now run into A in quantity equal to twice the bulk of the sugar, and allowed to stand, with frequent shaking, for ten or fifteen minutes. After this has been drawn off, a second and third portion, if necessary, is used, until the solution ceases to take up anything more, and the sugar under treatment has reached its maximum purity and whiteness. The washing with No. 1 solution is the most important in the process, and the time of washing and volume of wash-liquor employed must be left to the judgment of the operator, as they vary a great deal for different kinds of sugars. After the last portion of No. 1 has been carried off, successive quantities of Nos. 2, 3, and 4 are syphoned into A in the order named, and, after being shaken a few moments with the contents of the flask, removed. Finally the flask A is gently heated while the pump is still in operation, to facilitate the removal of the last traces of alcohol and ether. When the washings are finished the flasks are disconnected, the filtering-tube *o* (Fig. 35) taken out, carefully washed from any adhering particles of sugar into the flask by a wash-bottle, sufficient water added to dissolve the sugar together with a drop or two of lead solution, and the contents of the flask finally made up to 50 c.c., filtered, and polarized. The direct reading of the saccharimeter gives the percentage of crystallizable sugar.

The method of Payen–Scheibler, though apparently complicated, is in reality quite simple and easy of execution. Considerable care and some experience with it are, however, required to get good and unvarying results. The chief difficulty with the method—one which is especially prominent in a climate subject to such sudden changes as that of the United States—is that the solutions which at ordinary temperatures are saturated may become under or super-saturated, causing sometimes very serious errors unless constant care is taken. Even when the solutions are kept in bottles coated on the inside with sugar and almost filled with it, it has been known, in consequence of a sudden fall in the temperature of the laboratory, that the liquids, though not actually depositing in the storage bottles, were in a state of supersaturation, and as soon as a solid body was shaken with them, such as the raw sugar to be assayed, an immediate deposit of sugar was formed, sufficient to raise the test from 5 per cent. to 8 per cent. above the true amount. In case the solution is in this condition, or by a rise in temperature becomes capable of dissolving more cane-sugar, the difficulty may be surmounted by agitating briskly a portion of the solution for five minutes with a large excess of powdered sugar before using. It is important to observe, also, that the contents of the washing-flask A should remain at the same temperature, which should be the same as that of the solutions, throughout the operation; direct handling is therefore as much as possible to be avoided.

In consequence of the time necessary for a reliable determination by this method, or the misleading results of the estimation made in inexperienced or incompetent hands, the Payen–Scheibler method has never been generally used as a guide to the buyer of raw sugars, though it deserves to be.*

METHOD OF DUMAS.

M. Dumas found that alcohol of 85 per cent. by volume, containing fifty grammes of strong acetic acid to the litre, when saturated with cane-sugar, marked 74° on the alcoholometer of Gay-Lussac. For testing a sample of sugar, 100 c.c. of the normal liquor, prepared as above, is agitated with 50 grammes of the sugar to be tested, the solution filtered, and the areometer floated in it. If it marks 74°, the sugar contains 100 per cent. pure sugar; if it descends to 69°, the percentage is 95. Each degree lost by the areometer corresponds to one per cent. less in the titre of the sugar.

For sugars of 88 per cent. and upward this process may be made to give good results, but for lower products the

* Lotman (*Stammer's Jahresb.*, xiii. 156) has made an extensive series of analyses of raw beet and cane sugars, in which he compares the yield according to Scheibler's method with that by the *method of coefficients.* The results show that, with a few exceptions, Scheibler's yield is from .20 per cent. to 10.15 per cent. higher than by the latter method, the difference increasing as the sugars become lower in grade by a pretty even ratio.

results are unreliable. P. Casamajor* has proposed a modification of Dumas's method, which he highly recommends as giving good results on all classes of raw sugars except melados, or those containing a similar amount of water. He prepares the saturated alcoholic solution by agitating *methylic alcohol* of 83½° Tralles with powdered sugar. The solution, when fully saturated, marks 77.1° at 15° C. on the alcoholometer. The process for testing raw sugars is carried out as follows: 19.8 grammes of the sugar are weighed, well pulverized, and mixed in a mortar, as quickly as possible to avoid evaporation, with 50 c.c. of the standard solution; the mixture is poured upon a filter, and the density of the filtrate is taken with the areometer. To the degree of the alcoholometer, corrected for temperature, is added the difference between 100 and the alcoholometric degree of the standard solution. The sum represents the percentage of cane-sugar sought.

The readings of the Tralles instrument must be reduced to 15° C.; and to do this, for solutions between 60° and 70°, the number of degrees of temperature above 15° C. are multiplied by .37, and the product subtracted from the original reading of the areometer; for solutions above 70° the factor is .36, and for those below 60° the factor becomes .38. It is also advisable to make a correction on account of the volume of the normal solution used: At

15°	C.	19.8	grammes of	sugar	taken	for a vol. of	50	c.c.
20°	"	"	"	"	"	"	50.25	"
25°	"	"	"	"	"	"	50.5	"
30°	"	"	"	"	"	"	50.8	"
35°	"	"	"	"	"	"	51.2	"

* *Jour. of Amer. Chem. Soc.*, 1879, vol. i. No. 6.

For further details the reader is referred to the author's paper,* which states that the results obtained by this process, even on very low grade sugars, agree closely with duplicate assays made with the optical saccharimeter.

---

* *Chem. News*, xl. 74 *et seq.*

CHAPTER X.

ANALYSIS OF MOLASSES AND SYRUPS.

UNDER this head may be included all sugar solutions above a density of 15° or 20° Baumé, such as the brown and filtered liquors of the refinery, and the heavy syrups and molasses of the cane and beet sugar manufacture.

Estimation of the Cane-Sugar.—This estimation is made with the saccharimeter, as described under RAW SUGAR. The solutions should be weighed as quickly as possible to avoid evaporation. Molasses and impure syrups in general require a rather large quantity of lead solution and bone-black for decolorization. In some cases the ordinary method of procedure fails to give a solution light enough to admit of a saccharimetric reading, and it becomes necessary to either use the half-normal solution or the half-tube (100 mm.); the reading in either case must be multiplied by 2. When these means fail, it is best to proceed as follows: Weigh three times the normal quantity, dilute to 300 c.c. after adding lead solution, and filter. The solution, if still too dark, is submitted to a further filtration through a tube containing well-dried bone-black in grains, care being taken to reject the first third of the filtrate, as some sugar is retained by the char.*

In beet-sugar solutions there are generally impurities which affect the polarized ray sufficiently to cause the

* If the prepared black described on page 168 is used, the filtration with a tube is rarely or never necessary.

estimation of sugar with the saccharimeter to be more or less incorrect. These impurities are:

Malic acid.	Polarizing to	the left.
Aspartic acid (*in alkaline solution*).	"	"
Invert-sugar.	"	"
Metapectic acid.	"	"
Beet-gum.	"	"
Dextran.	"	the right.
Asparagine.	"	"
Aspartic acid (*in acid solution*).	"	"
Glutaminic acid.	"	"

The dextran and the beet-gum have a very high rotatory power.

Eissfeldt and Follenius * have published a process (for beet-juice) whereby these interfering impurities are either destroyed or precipitated, by warming the solution to be tested successively with alkaline solution of copper oxide containing a large excess of alkali, solutions of basic acetate of lead, and ferrocyanide of potassium, filtering, and polarizing. The results are said to be good.

Sickel † also weighs 13.024 grammes beet-juice, adds 1 c.c. of lead solution, and makes the liquid up to 50 c.c. with absolute alcohol, filters, and polarizes. The asparagine, aspartic acid, malic acid, gum, and dextran remain in the precipitate, while the presence of the alcohol neutralizes the rotatory effect of invert-sugar.

Tannic acid added to the warmed sugar solution precipi-

* *Zeit. f. Rubenz.*, 1877, 728. † *Ibid.*, 1877, 779.

tates many of the bodies which are optically active. When this agent is used, basic lead acetate, in quantity more than sufficient to precipitate all of the tannic acid, must be added after the latter. The tannic acid solution is prepared by dissolving 50 grammes of tannin in 200 c.c. of 90 per cent. alcohol, and diluting to one litre. On account of the large precipitate formed when tannin is used in connection with lead in very low products, the results are apt to be too high from the influence of the precipitate (page 166).* Clerget's method is hardly to be recommended to meet the difficulties in the case of optically-interfering impurities. Where the sugar is to be estimated with accuracy it will be advisable to have recourse to the method of inversion and estimation by Fehling's solution (estimation of cane-sugar, page 182). When the saccharine material is alkaline from the presence of caustic lime or alkalies, the solution should be barely acidified by acetic acid before the addition of the lead acetate, in order to neutralize the effect which alkalies have upon the polarized ray.

Estimation of Invert-Sugar.—As with raw sugar, page 217. The solution for titration must be dilute.

Estimation of Ash.—As with raw sugar, page 222. The solution, after the addition of sulphuric acid, ought to be heated cautiously, for fear of loss by spurting.

Estimation of the Water.—For purposes not requiring much accuracy this determination may be made with the Balling saccharometer, the reading indicating percentages of dry matter, which subtracted from 100 leaves the

---

* Champion and Pellet (*Sucrerie Indigène*, xii. 270) add 10 c.c. strong acetic acid to 100 c.c. of juice, or a proportional quantity to molasses, after the filtration from the lead precipitate in the ordinary process of decolorizing. This is said to completely neutralize the optical effect of asparagine.

water. For an accurate determination of water in sugar solutions, about twelve grammes of coarse, well-dried sand are weighed in a suitable dish or a watch-glass, together with a small rod and a glass bulb. This gives the first weight. Then allow from one to two grammes of the solution to drop into the bulb from a pipette, and reweigh. Finally break the bulb with a gentle pressure, taking care not to allow any fragment to fall from the dish, and carefully mix the syrup with the sand by means of the rod until a uniform mass is obtained. Dry at 100° C. for four or six hours. The bulbs can be easily blown over a common Bunsen lamp, and should have a small projecting neck and be thin enough to break easily. The diameter is about 12 mm. Example:

Weight of dish, sand, rod, bulb, and syrup...	22.121	grms.
" " " " "	20.104	"
Syrup taken..........................	2.017	"
The *ensemble* after drying 4 hours............	21.120	"
" " 6 "	21.119	"

$$22.121 - 21.119 = \frac{1.002 \times 100}{2.017} = 49.67 \text{ per cent. water.}$$

Estimation of Organic Matter not Sugar.—By difference, or one of the methods given under raw sugar.

Quotient of Purity or Exponent—*The Direct Method.*—The most direct, and in general the most convenient and reliable, way of obtaining this expression is to divide the percentage of impure sugar, or total solid matter, into the percentage of pure sugar multiplied by 100. The former is represented by the degree Balling of the solution reduced to standard temperature, while the latter is the polarization. The quotient expresses the percentage of pure sugar contained in the *dry substance—i.e.*, the total soluble matter if it were deprived of water.

Casamajor's Method.—This has the advantage of requiring no weighing; but where a balance is at hand the direct method is preferable both on account of absolute accuracy as well as the agreement of the results among themselves. According to the original method,* the solution to be tested is diluted so as to stand between 5° to 15° Brix; the degree Brix is taken, corrected for temperature, and the solution, after proper decolorization, is polarized as it stands, without weighing or dilution. The polarization is multiplied by a factor corresponding to the percentage of dry matter by Brix, the product being the quotient sought. The calculation may be made by the formulas—

$$Q = S \times \frac{16.19}{D \times P} \quad (1)$$

$$Q = S \times \frac{26.048}{D \times P} \quad (2)$$

in which

D is the degree Brix,
P the corresponding specific gravity,
S the polarization,

(1) is intended for use with instruments requiring the normal weight 16.19 grammes, and (2) with the Soleil-Ventzke.

I have found that the results given by this process approach nearer those of the *direct method*, and agree better among themselves, by having the solution less dilute than that given above; in this case the factor is decreased. The

* *Amer. Chemist*, vol. iv. 161.

following table, calculated by Mr. G. S. Eyster, of Boston, for use with this modification of the method, gives the factors by which the reading of the saccharimeter is to be increased, for the Soleil-Ventzke instrument. Example:

Reading of saccharimeter....... 50.1
Corrected degree Brix........... 25.7

Then from table—

$$\text{Factor for } 25.7^\circ = .914$$

$$50.1 \times .914 = 45.79 \text{ quotient.}$$

The solutions should be taken as strong as it is convenient, up to 27° Brix.

THE VERTICAL COLUMN SHOWS THE DEGREE BRIX IN WHOLE NUMBERS, WHILE THE HORIZONTAL COLUMN GIVES THE SAME IN TENTHS OF A DEGREE.

Degree Brix.	0	1	2	3	4	5	6	7	8	9	.01
15	1.6362	1.6248	1.6134	1.6024	1.5912	1.5801	1.5694	1.5587	1.5484	1.5379	.0011
16	1.5275	1.5175	1.5075	1.4979	1.4881	1.4783	1.4688	1.4594	1.4504	1.4414	.001
17	1.4320	1.4230	1.4141	1.4056	1.3969	1.3882	1.3798	1.3714	1.3632	1.3549	.0009
18	1.3467	1.3388	1.3310	1.3232	1.3157	1.3076	1.3001	1.2926	1.2853	1.2779	.0008
19	1.2706	1.2635	1.2564	1.2495	1.2425	1.2356	1.2288	1.2220	1.2155	1.2089	.0007
20	1.2020	1.1960	1.1880	1.1830	1.1770	1.1700	1.1640	1.1580	1.1520	1.1460	.0006
21	1.1400	1.1340	1.1280	1.1220	1.1170	1.1110	1.1060	1.1000	1.0940	1.0890	.0006
22	1.0830	1.0780	1.0730	1.0680	1.0630	1.0580	1.0530	1.0470	1.0420	1.0370	.0005
23	1.0330	1.0280	1.0220	1.0180	1.0130	1.0090	1.0040	.9990	.9950	.9900	.0005
24	.9850	.9810	.9760	.9720	.968	.963	.9590	.9550	.950	.9460	.0005
25	.9420	.9380	.9340	.9300	.926	.922	.9180	.9140	.910	.906	.0004
26	.9020	.8980	.8940	.8900	.887	.883	.8790	.8750	.872	.868	.0003
27	.8650	.8620	.8580	.8550	.851	.847	.8440	.8400	.837	.8330	.0003

The estimation of the quotient of purity is of great value in the various stages of the manufacturing and refining of sugar, showing as it does, in a single figure, the proportion of pure sugar to impurities. It is pre-eminently the test for the refiner, who in general does not wish to know *how much* sugar he has in a given solution, but *how pure* it is; its value for his purposes is altogether independent of the amount of water contained. To the buyer or seller, on the contrary, a knowledge of the percentage of pure sugar in the sample as it stands, otherwise known as the "direct test" or the "polarization," is of the greatest importance, and the percentage of water has a direct bearing upon it.

Calculation of the Results of an Analysis into Equivalents in the Dry Substance.—It is often desirable for purposes of comparison to have the results of an analysis reduced to terms of the dry substance. The reciprocal of the degree Brix multiplied by 100, is a factor by which the percentage of sugar, grape-sugar, and ash is increased to reduce them to the basis of dry mass; thus in a syrup having the following composition:

55° Brix.	
Cane-sugar....................	45.00
Grape-sugar...................	3.10
Ash	.82

the factor corresponding to 55 is 1.818; then in the dry substance we have:

Sugar.........45.00 × 1.818 =	81.81
Grape-sugar... 3.10 × 1.818 =	5.63
Organic matter by difference...	11.07
Ash........... .82 × 1.818 =	1.49
	100.00

These figures represent the quotients of purity respectively of the sugar, grape-sugar, organic matter, and ash. If the dry substance is estimated by drying, instead of the spindle, the results are, of course, more accurate. The table on page 193 may be used to obtain the factors necessary in the above calculation.

Estimation of the Color.—The determination of the color in relation to the sugar is made according to directions given under Raw Sugar, page 220. For sugar solutions, however, it is generally only necessary to estimate the color reduced to the normal standard of the colorimeter.

Estimation of the Alkalinity.—Products of the beet-sugar manufacture often contain caustic lime or alkalies. When these bodies are present in sufficient amount, it becomes necessary to determine them. For this purpose a standard alkaline solution is made by dissolving exactly 53 grammes of pure sodium carbonate, that has been heated some time to drive off moisture, in water, and diluting to one litre. A standard acid is also prepared by mixing 140 grammes of nitric acid, sp. gr. 1.385, or an equivalent amount of any other strength, with water, and diluting to about 1,100 c.c. The relation between the acid and alkali is now found by titration, using litmus or cochineal solution as an indicator. Suppose 20 c.c. of acid saturates 22 c.c. of alkali; then to make the acid solution *normal*—that is, to contain in one litre the number of grammes of the body dissolved corresponding to its combining weight (such solutions will consequently saturate each other volume for volume)—every 20 c.c. of acid must be diluted with two c.c. of water to bring it to the strength of the alkali, or for one litre $2 \times 50 = 100$ c.c. To one litre of the acid solution is added, accordingly, 100 c.c. of water, and the mixture well shaken.

Molasses and heavy syrups are often too much colored allow of the use of litmus or cochineal solutions, so that the point of saturation has to be determined with delicate litmus-paper.

Seventy-five grammes of molasses are weighed and diluted with water to 250 c.c.; two portions of 100 c.c. each, equivalent to 30 grammes of the molasses, are taken out for trial with the acid solution, the first to obtain an approximation of the alkalinity, and the second for a separate and more accurate determination. The alkalinity is generally calculated in terms of calcic oxide CaO.

1 c.c. of the test acid = .028 gramme CaO.

CHAPTER XI.

Analysis of the Cane and Cane-Juice.

THE CANE.

Estimation of Cane-Sugar.—It is difficult to obtain a sample faithfully representing the whole cane, as the amount of sugar differs in various parts. This variation is particularly marked at the joints. It is best to take three portions between the joints—from the base, top, and middle

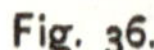

of the cane, together with one of the joints; slice the pieces and press out the juice in a small metallic roller-press (Fig. 36), moistening the pressed cane with hot water two or three times, and renewing the pressure to wash out the sugar contained as closely as possible. The juice from the press is diluted to the smallest number of cubic centimetres that it will be convenient to calculate upon. For example: Eight times the normal quantity for the Ventzke-Scheibler instrument, equal to 208.4 grammes of the cane, is weighed, pressed, and washed until about 380 c.c. of juice is obtained. This, for convenience, is diluted to 400 c.c. after the addition of lead solution, filtered, and polarized. If the polarization is 32, then, as eight times the normal was weighed, $\frac{32}{8} = 4$, which is multiplied by 4, on account of the dilution to 400 c.c. instead of 100 c.c., the standard volume gives 16. This is the percentage of cane-sugar in the sample treated.

The sugar may also be estimated by extraction with alcohol (page 180) on the dried assay.

The **grape-sugar** may be determined in a measured portion of the juice pressed from the cane, before the addition of lead solution or after its removal by sulphurous acid. See estimation of invert-sugar.

The *ash* and *water* are estimated in the manner already described in other places. It has only to be remarked that the slices should be made quite thin to ensure good drying, which is commenced at a temperature of 80°, and gradually raised to 110°.

CANE-JUICE.

The total impure sugar is estimated by the Brix saccharometer. Vivien's areometer (page 115) may also be found useful for this purpose. For hot countries, where cane-juice has in all cases to be tested, it is well to have hydrometers standardized at 25° C., instead of 15° or 17½°, as is the usual practice.

The Cane-Sugar is estimated by the saccharimeter, two or three times the normal quantity being weighed. The sugar can be more quickly determined, however, in cane-juice or any other weak saccharine solution by **Ventzke's process,** which dispenses with the weighing. This method is carried out by taking the density of the juice with the Brix spindle, finding the corresponding specific gravity from the table on page 116, and calculating the percentage of sugar according to the following formulas:

$$\frac{P \times .2605}{D} = S \quad (1)$$

$$\frac{P \times .1619}{D} = S \quad (2)$$

$$\frac{P \times .1500}{D} = S \quad (3)$$

in which P is the polarization of the juice as it stands without weighing; D = the density, and S the percentage of sugar. Formula (1) is for use with the Ventzke–Scheibler saccharimeter, (2) with the Soleil–Duboscq, and (3) for those instruments of which fifteen grammes is the normal weight. If the juice needs an addition of lead, it is filled into a 100 c.c. flask, and a measured volume of the lead solution added from a graduated pipette, the saccharimetric reading being increased in proportion to the dilution. Example: Cane-juice of 10° Brix (sp. gr. 1.04), to which 3 c.c. of lead solution to 100 c.c. were added, was found to polarize 32; 32 + 3 per cent. = 32.96. According to the formula (1)

$$\frac{32.96 \times .2605}{1.04} = 8.25 \text{ per cent. cane-sugar.}$$

A table is herewith given, reckoned according to formula (1) for the Soleil-Scheibler saccharimeter, which dispenses with the calculation. An example will show the manner of using it: A solution whose corrected per cent. of sugar by the Brix areometer is 9.5 polarizes 27; in the horizontal

column opposite 9.5, under 2, is found .502, which multiplied by 10 gives........................ 5.020
Under 7 in like manner occurs........................ 1.757

Percentage of cane-sugar........................ = 6.777

TABLE FOR ESTIMATING THE PERCENTAGE OF SUGAR BY WEIGHT, IN WEAK SUGAR SOLUTIONS: ABRIDGED FROM ONE CALCULATED BY OSWALD.

Degree Brix.	Sp. Gr.	Reading of the Saccharimeter.									
		1	2	3	4	5	6	7	8	9	10
0.	1.000	.260	.521	.781	1.042	1.302	1.563	1.823	2.084	2.344	2.605
.5	1.0019	.260	.520	.780	1.040	1.300	1.560	1.820	2.080	2.340	2.600
1.0	1.0039	.259	.519	.778	1.038	1.297	1.557	1.816	2.076	2.335	2.595
1.5	1.0058	.259	.518	.777	1.036	1.295	1.554	1.813	2.072	2.331	2.590
2.0	1.0078	.258	.517	.775	1.034	1.292	1.551	1.809	2.068	2.326	2.585
2.5	1.0097	.258	.516	.774	1.032	1.290	1.548	1.806	2.064	2.322	2.580
3.0	1.0117	.257	.515	.772	1.029	1.287	1.545	1.802	2.060	2.317	2.575
3.5	1.0137	.257	.514	.771	1.028	1.285	1.542	1.799	2.056	2.313	2.570
4.0	1.0157	.256	.513	.769	1.026	1.282	1.539	1.795	2.052	2.308	2.565
4.5	1.0177	.256	.512	.768	1.024	1.280	1.536	1.792	2.048	2.304	2.559
5.0	1.0197	.255	.511	.766	1.023	1.277	1.533	1.788	2.044	2.299	2.554
5.5	1.0213	.255	.510	.765	1.020	1.275	1.530	1.785	2.040	2.295	2.549
6.0	1.0237	.254	.509	.763	1.018	1.272	1.527	1.781	2.036	2.290	2.544
6.5	1.0257	.254	.508	.762	1.016	1.270	1.524	1.778	2.032	2.285	2.539
7.0	1.0278	.253	.507	.760	1.014	1.267	1.521	1.774	2.027	2.281	2.534
7.5	1.0298	.253	.506	.758	1.012	1.265	1.518	1.771	2.023	2.276	2.529
8.0	1.0319	.252	.505	.757	1.010	1.262	1.515	1.767	2.019	2.272	2.524
8.5	1.0339	.252	.504	.756	1.008	1.260	1.512	1.763	2.015	2.267	2.519
9.0	1.0360	.251	.503	.754	1.006	1.257	1.509	1.760	2.011	2.263	2.514
9.5	1.0380	.251	.502	.753	1.004	1.255	1.506	1.757	2.007	2.258	2.509
10.0	1.0410	.250	.501	.751	1.002	1.252	1.503	1.753	2.003	2.254	2.504
10.5	1.0422	.250	.500	.750	1.000	1.250	1.500	1.750	1.999	2.249	2.499
11.0	1.0443	.249	.499	.748	.998	1.247	1.497	1.746	1.995	2.245	2.494
11.5	1.0464	.249	.498	.747	.996	1.245	1.494	1.743	1.991	2.240	2.489
12.0	1.0485	.248	.497	.745	.994	1.242	1.491	1.739	1.987	2.236	2.484
12.5	1.0506	.248	.496	.744	.992	1.240	1.488	1.735	1.983	2.231	2.479
13.0	1.0528	.247	.495	.742	.990	1.237	1.484	1.732	1.979	2.227	2.474
13.5	1.0549	.247	.494	.741	.988	1.235	1.482	1.728	1.975	2.222	2.469
14.0	1.0570	.246	.493	.739	.986	1.232	1.479	1.725	1.971	2.218	2.464
14.5	1.0591	.246	.492	.738	.984	1.230	1.476	1.722	1.967	2.213	2.459
15.0	1.0613	.245	.491	.736	.982	1.227	1.473	1.718	1.963	2.209	2.454
15.5	1.0635	.245	.490	.735	.980	1.225	1.470	1.714	1.959	2.204	2.449
16.0	1.0657	.244	.489	.733	.978	1.222	1.467	1.711	1.955	2.200	2.444
16.5	1.0678	.244	.488	.732	.976	1.220	1.464	1.708	1.951	2.195	2.439
17.0	1.0700	.243	.487	.730	.974	1.217	1.461	1.704	1.948	2.191	2.434
17.5	1.0722	.243	.486	.729	.972	1.215	1.458	1.701	1.944	2.186	2.429
18.0	1.0744	.242	.485	.727	.970	1.212	1.455	1.697	1.940	2.182	2.424
18.5	1.0765	.242	.484	.726	.968	1.210	1.452	1.694	1.936	2.178	2.420
19.0	1.0787	.241	.483	.724	.966	1.207	1.449	1.690	1.932	2.173	2.415
19.5	1.0810	.241	.482	.723	.964	1.205	1.446	1.687	1.928	2.169	2.410
20.0	1.0833	.240	.481	.721	.962	1.202	1.443	1.683	1.924	2.164	2.405
20.5	1.0855	.240	.480	.720	.960	1.200	1.440	1.680	1.920	2.160	2.400
21.0	1.0878	.239	.479	.718	.958	1.197	1.437	1.676	1.916	2.155	2.395
21.5	1.0900	.239	.478	.717	.956	1.195	1.434	1.673	1.912	2.151	2.390
22.0	1.0923	.238	.477	.715	.954	1.192	1.431	1.669	1.908	2.146	2.385
22.5	1.0946	.238	.476	.714	.952	1.190	1.428	1.666	1.904	2.142	2.380
23.0	1.0969	.237	.475	.712	.950	1.187	1.425	1.662	1.900	2.137	2.375

Grape-Sugar.—(See chapter x.)

Ash.—(See chapter x.)

Estimation of Water.—By the Brix spindle or by drying, according to the accuracy required (page 252).

The *quotient of purity* is a very useful determination, and may be made by the direct method:

$$\frac{\text{Pol.} \times 100}{\text{Brix}} = \text{Exp.}$$

CHAPTER XII.

Analysis of the Beet and Beet-Juice.

THE BEET.

Preparation of the Sample.—The top and small radicles are cut off, and the beet is washed to free from mechanical impurities, being dried with a coarse towel. If desired, the weight before and after this treatment may be taken. If a single beet is to be operated upon, the whole, after the above preparation, is reduced to a fine pulp by grating or any other means. For a sample representing a quantity of beets or the growth of a field, it is necessary to take a number of roots differing in size and variety. Beets taken from the same field, and apparently submitted to the same conditions, are found to vary a great deal in their saccharine richness.

By successive slices, made parallel to the axis of the beet, cut out a square prism, of such thickness as will be determined by the size of the roots. Fig. 37 will illustrate this. The dotted line represents in projection the prism. The different portions thus obtained from all the roots constituting the average are reduced to pulp and mixed together. Champonnois has invented a boring *râpe* which serves very well to cut out a portion of the beet as above described, and which pulps it at the same time (Fig. 38).

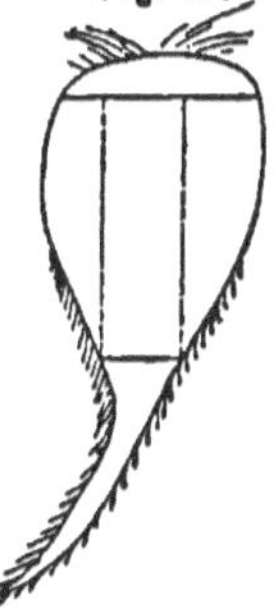
Fig. 37.

Estimation of Cane-Sugar.—About 200 grammes of the pulp are placed in a small filtering-bag and pressed in a hand-press slowly until the juice ceases to flow with the strongest pressure obtainable; the marc is then moistened with boiling water, the pressure renewed, and this operation repeated until all soluble matter has been extracted and the residue is dry, care being taken to avoid undue dilution of the solution. Three to four cubic centimetres of tannin solution are added, and about three times that volume

Fig. 38.

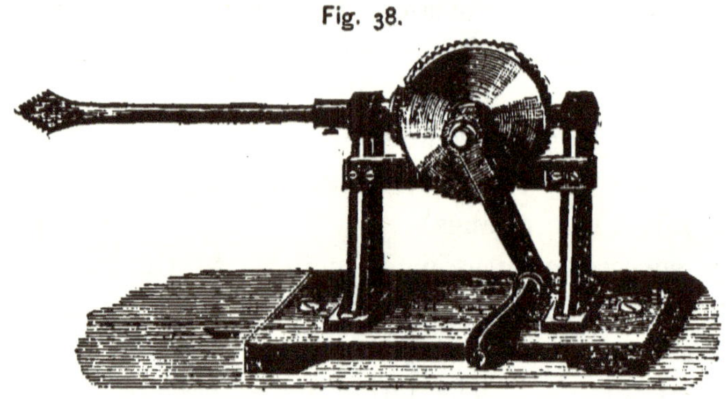

of lead solution, and the whole made up to an exact volume, filtered, and a portion polarized. Any of the modifications suggested in the case of molasses (page 251) may be tried. The calculation is in all respects similar to that for estimating the sugar in the cane.

Scheibler's Method for estimating Sugar in the Beet.*—The estimation, as given above, will show only approximately the amount of sugar, on account of impurities present in the juice obtained, which have a considerable effect upon the polarized ray—as well as the generally imperfect extraction of the sugar. Scheibler's process,

* German Patent, No. 3573, *Stammer's Jahrb.*, 1879.

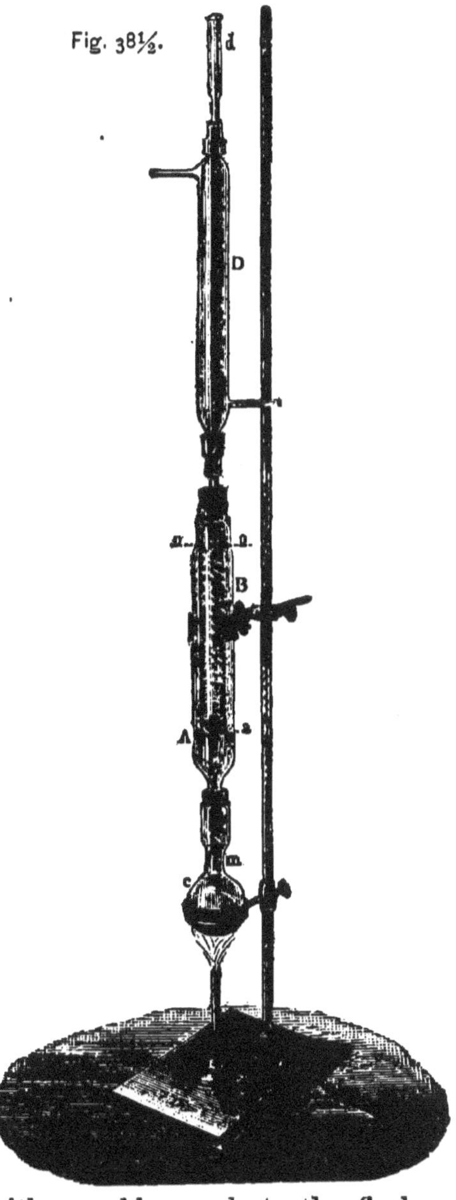

Fig. 38½.

though requiring special apparatus, avoids these errors, and gives results very near the truth. It is essentially an extraction of the sugar from the finely-divided beet without previous drying, by means of alcohol, in the heat.

The apparatus is shown in Fig. 38½. D is an upright glass condenser fitted tightly to A by a rubber cork. A and B are glass tubes of the form shown, A being placed within B, and making a close joint at the upper ends, the lower portion of A being free and open. Near the top of the latter are two or more openings, five to six mm. in diameter, *o o*, communicating between the annular space and the interior of A. B is fastened with a rubber cork to the flask *c*, which is graduated to 50 c.c. The narrowed end of B extends some distance into *c*, in order that none of the boiling sugar solution may be spurted into the former. In the cylinder A is placed a small filter, *a*, which may be of cotton, felt, or other suitable material.

For the execution of the test from 20 to 25 grammes beet-pulp are placed in A, which has been previously tared, by means of a long-stemmed funnel, so that the mass fills the tube nearly to *o o*. A is then reweighed, and the difference of the two weighings gives the amount of assay taken. The cylinders A and B are now adjusted as shown in the cut, and the condenser fixed in position. Now 25 c.c. alcohol of 90 to 94 per cent. Tralles (.8339 to .820 sp. gr.) are placed in the flask, and by means of a sand or water bath heated to boiling. It is perhaps better that the alcohol should be added through the top of the condenser at *d*, through which it passes to the beet-pulp and falls in *c*. The vapor from the boiling alcohol, ascending into the space between A and B, passes through *o o*, and is liquefied in the condenser (kept cool by a stream of water), from which it drops in A, and, coming in contact with the assay, extracts the sugar, the solution dropping into the flask, where it parts with its alcohol, which is again made to pass through the substance to be extracted. The flame should be so regulated that the drops from A should succeed each other at regular intervals, and not too quickly. As a rule, from a half to three-quarters of an hour's boiling is sufficient to complete the operation, after which the apparatus is allowed to cool, the last drops of the solution from A being received in the flask, which is filled to the mark with distilled water, after the addition of a few drops of lead, filtered, and the sugar estimated with the saccharimeter. The alcoholic sugar solution, after dilution to 50 c.c., should show about 41 per cent. Tralles. For many varieties of beets the strengths given for the alcohol cannot be strictly adhered to, as when the latter is too dilute a troublesome frothing takes place on boiling.

According to Scheibler and Tollens (*loc. cit.*), the continued boiling of the alcoholic solution causes no sensible alteration of the sugar dissolved, even when the beets operated upon are slightly acid. The method has been exhaustively and critically examined by Tollens * and others, with the result of establishing its substantial accuracy as showing the absolute amount of sugar in the beet, and its great superiority over processes previously in use.

Estimation of the Marc and the Amount of Juice.—Twenty grammes of the pulp are made into a thin paste with boiling water, poured upon a weighed filter, and thoroughly washed, with the aid of a vacuum if necessary. The filter is then dried at 110°. Example:

Watch-glasses + filter + marc at 110° =	22.100
" + " at 110° =	21.260
Weight of marc......................	.840

$$\frac{.840 \times 100}{20} = 4.2 \text{ per cent.}$$

The percentage of marc subtracted from 100 gives the percentage of juice, as

$$100 - 4.2 = 95.80 \text{ per cent. juice.}$$

The amount of juice may be obtained by an indirect method which gives results agreeing very well with the above. The water is determined in the pulp by drying,

* *Zeit. f. Rubenz.*, May, 1880; *Stammer's Lehrbuch*, Ergänzungsband, 102.

and also in the juice; then the percentage of juice is found by the formula $\frac{s}{S} \times 100$, in which s is the percentage of water in the pulp, and S that in the juice. The marc is obtained by difference.

Scheibler's Method.—The percentage of marc, and from it that of the juice, may be obtained with greater accuracy than by the methods described, in connection with Scheibler's process for determining the sugar in the beet (page 266). The contents of the tube A, after the extraction of the sugar, are desiccated by passing a stream of dry air through the latter, after which it is weighed and the amount of marc calculated. Scheibler claims that the results obtained by the formula $\frac{s}{S} \times 100$ are erroneous, as the direct polarization of the juice is never quite correct, owing to the presence of about five per cent. of sugar-free water in the beet (*colloidal water*).

Grape-Sugar is generally present in very small quantities. To estimate it a weighed portion of the pulp is extracted with water, and the grape-sugar determined in the expressed liquor by Fehling's method, with the usual precautions (see estimation of dextrose).

Water is determined by drying the pulp or the thinly-sliced beet at 100°.

The Estimation of Ash.—As in raw sugar (page 222).

ANALYSIS OF BEET-JUICE.

The Baumé hydrometer is largely used to afford a relative comparison as to the value of beet-juice. The Brix spindle is, however, preferable, in that the readings correspond, within certain limits of error, to the percentage of impure sugar in solution.

Estimation of Cane-Sugar.—By the saccharimeter, twice or thrice the normal quantity being taken and diluted up to 100 c.c., after the addition of about 2 c.c. tannin and 6 c.c. of lead solutions. The cane-sugar may also be readily determined by Ventzke's method (page 261). As with beet-molasses, though in a less degree, this estimation is rendered more or less incorrect by the presence of optically active impurities in the juice. For modifications of the usual method to be pursued to meet this source of error, see the chapter on the analysis of molasses (page 250).

Estimation of Grape-Sugar.—As in the case of the beet.

Estimation of the Ash.—With sulphuric acid, as with the beet. The juice should be carefully evaporated to avoid loss, before the charring takes place.

Estimation of Water.—By drying in sand with bulb (page 252) for accurate work, by preference *in vacuo*—or with the Balling spindle.

Quotient of Purity.—Divide the degree Balling, corrected for temperature, into the percentage of cane-sugar by polarization × 100. Stammer gives as the valuation-coefficient (*Werthzahl*) of beet-juice, an expression obtained by multiplying the quotient into the percentage of cane-sugar, and dividing by 100. This is only useful for comparative purposes.

Estimation of Organic Matter not Sugar.—This is determined by difference or any of the methods given in chapter ix. Where the water is estimated by the areometer the results are always low, owing to the error of the instrument in impure solutions, and consequently the matters determined by difference are too high. To correct this error, Stammer has proposed to subtract one-fifth of the organic matter thus found, and add it to the water. This correction would equally apply to all impure sugar solutions, whether from the beet or cane.

Estimation of the Alkalinity.—On the juice without dilution (page 258).

Estimation of the Color.—As with raw sugar and molasses.

The wet analysis of beet-juice may be reduced to dry substance, as shown on page 257.

Note.—The matter given in relation to the analysis of cane and beet juice applies equally to any weak sugar solution, such as the "sweet water" from char-washing, etc.

CHAPTER XIII.

Analysis of Waste Products.

ANALYSIS OF SCUMS AND SOLID RESIDUES.

THESE consist of the refinery scums; the marc of the beet freed from all obtainable sugar; the bagass, or residue from the cane-presses; and the precipitates produced in the process of carbonatation and defecation in the beet-sugar manufacture. The only estimation commonly made upon these bodies is that of the sugar. Before these residues are thrown out in the course of the manufacture, it is of considerable importance to make sure that there is no undue proportion of sugar present. They should be tested systematically, and sufficiently often to form a proper control of the work.

Refinery Scum.—This is the matter caught in the bag-filters when the crude solution of raw sugar is filtered preparatory to being run upon the char. It consists of the insoluble matters contained in the raw sugar, as sand, foreign matters of all kinds, particles of cane-fibre, the substances precipitated by caustic lime in defecation, and the coagulated albumen and bodies carried down with it, when blood is used in the process of defecation.

Estimation of Cane-Sugar.—From a large average sample, a smaller one is prepared by taking out portions and thoroughly mixing them together. Weigh 13.04 grammes for the Ventzke-Scheibler, or the normal quantity for other saccharimeters, add enough boiling water to make a uniform paste, and gradually dilute with the hot water until the weighing-capsule is nearly full and a uniform thin magma is obtained free from lumps; pour this upon a filter in a funnel provided with a filtering-cone, and filter by the aid of a vacuum into a flask or cylinder graduated to 100 c.c., until all the liquid has passed through; add small portions of boiling water at a time, stirring up the insoluble matter on the filter as much as possible with the stream from the wash-bottle, and continue the washing until the filtrate measures nearly 100 c.c.; if the solution is alkaline, barely acidify with acetic acid, add a few drops of lead solution, allow to cool, fill to the mark, shake, add a little powdered bone-black, filter, and polarize. The reading (by the Ventzke-Scheibler instrument), multiplied by two, gives the percentage of cane-sugar. This method is accurate enough for nearly all purposes; but where greater exactness is required the scum may be extracted with a larger quantity of hot water, and the sugar determined in an aliquot part of the filtrate after inversion, by Fehling's method.

If the *grape-sugar* is to be determined, the solution is made to 100 c.c. before the addition of the lead, and an aliquot part of it taken for the grape-sugar estimation. From the remainder a 50 c.c. flask is filled, a measured volume of lead solution added, the solution filtered and polarized. The reading must be corrected for the dilution caused by the addition of the lead.

The *water* is determined by drying one gramme at 100° to 110° C.

The *ash* is determined by incineration without the addition of sulphuric acid.

Beet Marc.—The cane-sugar may be determined in the same manner as with refinery scums, or better after Scheibler's method with alcohol (page 266). If the residue is very poor in sugar, it would be advisable to estimate the latter by the inversion with hydrochloric acid, and Fehling's method, after extraction with a large quantity of hot water. The other estimations may be made as in the case of refinery scums.

Residues from the Carbonatation Process.—These form the precipitates produced by adding a large excess of caustic lime to the sugar solutions, and precipitating the solution of calcic sucrate with a stream of carbonic acid gas. They are frequently alkaline from imperfect carbonatation, and the sugar contained is in the state of sucrate. The estimation of the *cane-sugar* may be made similarly to that of refinery scums, except that it is necessary to first diffuse the solid matter through water, and pass a stream of washed carbonic acid gas to break up the combination of the sugar with the lime ; filter from the precipitated calcium carbonate, and determine the sugar in the filtrate by the saccharimeter, or with alkaline copper solution after inversion.

E. Perrott * gives a method that is equally applicable to the determination of cane-sugar in all sucrates. One hundred grammes of substance are taken, mixed well with 380 c.c. of water, at the same time breaking up all lumps, and 20 c.c. of carbonate of ammonia solution. The mixture is allowed to stand ten minutes after agitation, and filtered. From the filtrate 200 c.c., representing 50 grammes of assay, are taken, diluted to 400 c.c., and the cane-sugar determined by Fehling's method after inversion.

---

* *Sucrerie Indigène*, ix. 11.

Bagass.—Two hundred grammes are reduced to as fine a state of division as possible, mixed well with boiling water, placed in a small filtering-bag, and pressed with a hand-press. The washing is repeated with fresh portions of water, and the pressing renewed until all sugar is extracted. As the solution is commonly too dilute to test to advantage with the polariscope, it is best to take an aliquot part of that obtained, corresponding to a known weight of bagass, and to estimate the grape-sugar directly, and the cane-sugar after inversion, by Fehling's method.

WASTE WATERS.

Under this head are included the last washings of the bag and char filters, and those of the diffusion and maceration processes of the beet-sugar manufacture. It is especially important to know when the washings no longer contain enough sugar to make it advantageous to save them. Ten c.c. of the waste waters are evaporated at a water-bath heat in tared dishes, and the net residue represents the amount of solid matter contained, of which from twenty to seventy-five per cent. may be sugar. If it is wished to estimate the amount of sugar, a larger portion is evaporated with the addition of a few drops of hydrochloric acid, and the amount of invert-sugar determined by Fehling's method.

Estimation of Cane-Sugar in Dilute Solutions.—In testing very dilute solutions for sugar the following method of procedure may be adopted: Evaporate the solution after careful neutralization, if necessary, to from one-fifth to one-twentieth of its bulk, on a water-bath at a low heat, and determine the grape-sugar directly, and the cane-sugar after inversion by Fehling's method. As a rule, for very dilute solutions, the mere presence of sugar of any kind is sought to be demonstrated, so that it is only necessary to evaporate with the addition of a little hydrochloric acid, and determine the invert-sugar found.

CHAPTER XIV.

ANALYSIS OF COMMERCIAL GLUCOSE OR STARCH-SUGAR.

Grape-Sugar — Corn-Sugar.

Starkezucker, Kornzucker, Gr.—Sucre de Fécule, Fr.

THIS product is prepared from corn-meal or starch, either by the action of mineral acids at a boiling temperature, or by means of diastase. It occurs commercially in three forms—viz., in the condition of a dry granular or fine powder; as a solid in lumps containing varying amounts of water; and as a thick yellowish or white syrup. The following analyses will show the composition of different varieties:

I. By Steiner.*

	I.	II.	III.	IV.
Water	15 50	6.00	13.30	7.60
Ash	.30	2.50	.40	1.10
Dextrose	45.40	26.50	76.00	....
Maltose	28.00	40.30	5.00	42.60
Dextrin	9.30	15.90	....	39.80
Carbohydrates	1.50	7.00	5.30	8.90
Protein substances	traces.	1.80	.20	....
Acid (as SO_4H_2)	.08	.03	.05	....
Iodine reaction	....	distinctly blue.	....	....
	100.08	100.03	100.25	100.00

* *Stammer's Jahrb.*, 1879, 379; Dingler, ccxxxiii. 262.

The first is of German origin, white and soft; the rest are English, produced by the action of dilute sulphuric acid on corn-meal at high pressure.

II.

	Powdered.	Granulated.	Lumps.	Syrup.
Sugar by copper test..	81.63	74.27	71.26	50
Undetermined bodies	9.06	11.89	12.57	30
Water...............	8.76	13.34	15.71	19
Ash.................	.55	.50	.46	1
	100.00	100.00	100.00	100.00

In III., the next series of analyses, by Neubauer, the sugar is estimated by the fermentation method:

	I.	II.	III.	IV.
Fermentable sugar....................	57.20	63.02	61.43	59.25
Non-ferment bodies (dextrin, etc.)......	18.38	13.32	22.45	23.59
Water................................	24.42	23.66	16.12	17.16
	100.00	100.00	100.00	100.00

The non-fermentable, or bodies classed as undetermined, consist of dextrin, unaltered starch, and, according to Haarstick,* of the *amylin* of Béchamp. They have a high dextro-rotation.

The solid varieties of commercial glucose show *birotation* in a marked degree, while with the syrups this property is generally absent. The latter differ from the former in that the conversion of the starch into sugar is not carried so

* *Stammer's Jahrb.*, 1876, 176.

far, and hence the amount of organic matter not sugar in them is proportionately large.

ESTIMATION OF THE SUGAR BY FEHLING'S METHOD.

On account of the presence of maltose with the dextrose, sometimes in large amounts as shown by Steiner's results, this determination cannot show anything definite as representing dextrose. The amount of copper oxide reduced by the two sugars differs very much, 100 parts of maltose reducing 141.5 parts CuO, while the same quantity of dextrose throws down 220 parts CuO. The results of this test have accordingly only a relative value. As to the action of dextrin upon the heated copper liquor, Rumpf and Heinzerling,* as the result of their investigation, state that solutions of (1) caustic soda and cupric sulphate at the boiling-point do not act on dextrin entirely free from sugar, which corrects Gerhardt's observation, who asserted that dextrin caused a reduction of the oxide in the sulphate; (2) solutions of alkaline tartrates, and Fehling's solution each act upon dextrin, making the results of the dextrose estimation too high in direct proportion to the length of time the heating is continued. When the reduction is quickly effected, and the heating continues only a few minutes, they have found that the error in the estimation of dextrose in the presence of dextrin in starch-sugars is too small to sensibly affect the results.

The execution of the test is in all respects according to directions already given. See chapter viii.

---

* *Zeit. f. Anal. Chemie*, ix. 358.

ESTIMATION OF THE DEXTROSE BY FERMENTATION.*

A solution of the sugar to be examined is made containing a known amount, and the percentage of dry matter determined. The solution is then submitted to fermentation with yeast, and, after the expulsion of the alcohol and carbonic acid formed, the percentage of dry matter is again determined, and the difference between the amounts of dry substance estimated before and after fermentation gives the sugar originally present. The results are a little low as compared with those given by Fehling's method, because in the vinous fermentation all of the sugar does not break up into alcohol and carbonic acid, but about five per cent. is converted into glycerin, succinic acid, and other bodies, which, being non-volatile at the temperature of boiling water, remain in the liquid after the evaporation.

Example: One hundred grammes of starch-sugar are dissolved, diluted to one litre, and the specific gravity taken. Suppose it is 1.03: we find from Balling's table (page 116) that this corresponds to a percentage of dry substance of 7.463, and as 100 c.c. weigh 103 grammes, 100 grammes of the solution contain 9.708 grammes of the original assay, and

$$9.708 : 7.463 :: 100 : x = 76.87 \text{ per cent.}$$

The composition of the solution, then, is

76.87 per cent. dry substance.
23.13 " water.
100.00

For the fermentation 500 c.c. of this solution are taken,

* Neubauer, *Wagner's Jahresb.*, 1875, 806.

a sufficient quantity of fresh beer-yeast added, and the whole placed in a fermentation apparatus arranged so that dried carbonic acid can escape. Compare matter on page 181. The system is then weighed, and allowed to remain at the proper temperature for three or four days until the action is complete. This point may be ascertained by weighing at intervals. When the apparatus ceases to lose weight the operation may be considered as finished. The liquid in the flask is filtered, boiled down to one-third of its volume to drive off alcohol, and, after cooling, made up to its original bulk. If the density after fermentation is 1.0082, which corresponds to 2.05 per cent. dry matter, 100 c.c. weigh 100.82 grammes and contain 2.068 grammes dry substance; or in 500 c.c., 10.340 grammes. As the 500 c.c. of solution contained 50 grammes of the original sugar, then $\frac{10.340 \times 100}{50} = 20.67$ per cent. unfermentable matter; $76.87 - 20.67 = 56.20$ per cent. of fermentable sugar.

ANTHON'S METHOD.

This is based on the fact that the impurities present in commercial starch-sugar have a greater density than that of the sugar contained. The process, though somewhat empirical, is said to give results accurate enough for most purposes.

A saturated solution of the sugar to be examined is made by adding a large excess of it, in as fine a state of division as possible, to water, and allowing the mixture to stand, with frequent agitation, for twelve hours, or until fully saturated. The specific gravity of the clear solution thus produced is obtained either by the specific-gravity balance or by weighing (chapter v.) From this the percentage of impurities may be found in the accompanying table:

TABLE.

Density of sat. solution.	Per ct. of impurities.	Density of sat. solution.	Per ct. of impurities.	Density of sat. solution.	Per ct. of impurities.
1.2060	0	1.2350	15	1.2587	30
1.2082	1	1.2368	16	1.2603	31
1.2104	2	1.2386	17	1.2618	32
1.2125	3	1.2404	18	1.2633	33
1.2147	4	1.2422	19	1.2649	34
1.2169	5	1.2440	20	1.2665	35
1.2189	6	1.2456	21	1.2680	36
1.2208	7	1.2473	22	1.2695	37
1.2228	8	1.2489	23	1.2710	38
1.2247	9	1.2506	24	1.2725	39
1.2267	10	1.2522	25	1.2740	40
1.2284	11	1.2535	26	1.2755	41
1.2300	12	1.2548	27	1.2770	42
1.2317	13	1.2561	28	1.2785	43
1.2333	14	1.2574	29		

ESTIMATION OF THE WATER.

Two to three grammes are weighed and dried with sand (page 219). In the case of solid glucose, the portion to be tested is placed on the weighing-dish, separated from the sand, and melted with a gentle heat. When liquefied it is mixed with the sand in the usual manner.

The dextrin and other matters are estimated by difference, after the ash is determined by incineration, with the addition of sulphuric acid.*

* *Estimation of the Dextrose optically.*—This determination cannot be made by the optical method, on account of the presence of a large and variable amount of dextrin, maltose, and other bodies, which are optically active, and whose specific rotatory powers are different from, and much greater than, that of dextrose. The specific rotatory power of dextrin varies from $[\alpha]_j = 139°$ to 212°; while that of maltose is $[\alpha]_D = 139.8°$. If it is desired, for purposes of comparison, to polarize starch-sugar, the solution before it is placed in the tube of the saccharimeter for observation, should be heated for five minutes to 100° to get rid of the birotation, and obtain at once the lowest reading.

THE DETECTION OF DEXTRIN AND STARCH-SUGAR WHEN MIXED WITH RAW AND REFINED SUGARS.

I. **The Adulteration of Raw Sugar with Dextrin.** —Commercial dextrin has been added to raw sugars in order to give them a higher polarization, and consequently a greater market value; .40 of one per cent. of dextrin raises the saccharimetric titre about one per cent. Two qualitative tests are commonly resorted to for detecting dextrin under these conditions, though neither is entirely reliable: 1. Alcohol of 95 per cent. added to a concentrated solution of sugar containing the adulterant gives a white, thread-like coagulum, while more dilute solutions show only a cloudiness in a greater or less degree. The salts present in raw sugar, and particularly sulphate of lime, give a similar precipitate. 2. A solution of iodine in iodide of potassium produces with dextrin, according to the method of manufacture, a wine or violet red, while some varieties do not give any coloration.* The presence of dextrin may be detected with certainty by Chandler and Rickett's method (page 287). For the determination of cane-sugar the process of inversion and estimation with copper liquor will have to be resorted to (chap. viii.) †

II. *Detection of Starch Sugar or Syrup when mixed with Raw or Refined Sugars.*—The presence of these sub-

---

* Boivin and Loiseau (*Wagner's Jahresb.*, 1870, 399) give the following as the marks of sugars containing dextrin: 1. On burning they give off the odor of heated bread. 2. They are very difficult to filter, and the filtrates are apt to be cloudy. This is particularly the case when lead solution has been used in clarifying. 3. Owing to imperfect mixture, separate lumps of dextrin may be separated and appropriately tested.

† Lactose or milk-sugar in raw sugar may be detected by treating the latter with twelve times its weight of 80 per cent. alcohol, which dissolves the sugar and leaves the lactose.

stances may in general be shown by paying attention to the following points: 1. Sugars mixed with powdered or granulated corn glucose, on solution in water invariably leave white particles of the glucose undissolved; 2. Owing to the birotation exhibited by solid starch and corn glucose, it will be observed, on submitting a commercial sugar containing it to the polariscopic test, that the reading does not remain constant, but gradually becomes less until a point is reached when the diminution of the reading ceases. If the solution is observed immediately after it is prepared (without heat), as little as three to five per cent. of starch-sugar may be thus detected. This test only applies when the sugar is mixed with solid glucose, as the syrup does not show birotation. 3. On account of the high rotatory power of starch-sugar, a refined sugar mixed with it will show a larger percentage of cane-sugar by the saccharimeter than the true one; hence the analysis generally adds up over one hundred. This will apply whether the material used for mixing is solid glucose or the syrup.

With these three tests it is easy to determine qualitatively the presence of starch or corn glucose in any sample of sugar, whether raw or refined, in amounts from two per cent. upwards.

There exists no accurate method for determining the amount of commercial glucose in any refined or raw sugar mixed with it. The glucose itself varies greatly in composition, and the invert-sugar contained in raw and refined sugars acts toward Fehling's solution precisely as does the sugar in glucose. The ordinary optical method cannot be employed, because the reading of the saccharimeter given by a mixture of cane and starch sugars is a resultant of the rotations of the two sugars, together with that of the impurities present in the latter. The rotation of starch-sugars from different sources and in different conditions, whether solid or liquid, varies within exceedingly wide limits. Clerget's method is equally inapplicable, except as a qualitative test, for the reasons stated above. The unsuitableness of this method for the quantitative estimation is specially prominent on account of the optical properties of the maltose, dextrose, dextrin, and soluble starch present, it being remembered (page 136) that Clerget's process is intended for solutions of cane-sugar containing no rotatory substance other than optically active invert-sugar of known specific rotatory power.

Casamajor * recommends the use of methylic alcohol marking 50° of Gay Lussac's alcoholometer, saturated with starch-sugar, as a qualitative test for the latter when mixed with commercial cane-sugars. The suspected sugar, after drying, is thoroughly washed with the test solution, which dissolves the cane-sugar and impurities, leaving the glucose in grains and powder. It seems probable, as the author suggests, that this method might be so modified as to give fairly good results quantitatively, perhaps better than with the very unsatisfactory methods hitherto proposed, by collecting the undissolved starch-sugar on a weighed filter, after all soluble matters have been removed by the alcoholic sugar solution, and the strongest methylic alcohol (92½° Gay Lussac), applied successively.

Drs. Chandler and Ricketts † have devised a method for estimating the right-rotating substances in the glucose added to a commercial sugar.

---

* *Jour. Am. Chem. Soc.*, ii. 428.

† *Ibid.*, vol. i.

CHANDLER AND RICKETTS'S METHOD.

This consists in inverting the mixed sugars with acids, as in Clerget's process (page 137), and observing the rotation in a water-bath tube at 92° C. (temperature of water-bath). Invert-sugar at 87.2° C. has no effect upon the polarized ray, owing to the fact that the rotation of levulose is neutralized by that of the dextrose which is constant for all temperatures (see invert-sugar, page 89). Hence, when a mixed sugar of commerce is inverted, the cane-sugar is converted into invert-sugar, which, with that originally present, is optically inactive at the temperature named. The dextrose and other bodies from the starch-sugar preserve their specific rotatory effect. When, therefore, a pure commercial sugar is inverted, at 87.2° the rotation is null, while if any corn glucose is present a rotation to the right will be shown, and in proportion to the amount present.

To calculate the results given by this process a standard starch-sugar was taken which gave "an average rotation to the right at 92° C. of 87 divisions of the saccharimeter scale (Ventzke-Scheibler), when the sample tested 63 per cent. by Fehling's method. Hence, if 26.048 grammes be the amount taken for observation, $\frac{26.048 \times \frac{63}{100}}{87 \times 100} = 18.864$ grammes is the amount of dry substance necessary to read 100 divisions on the scale, or each division is equal to .1886 gramme." 26.048 grammes of the suspected sugar is taken for the Ventzke-Scheibler instrument, inverted with hydrochloric acid, and the solution observed in the tube heated to 92° C. Each division of the scale read corresponds to .1886 gramme reducing substances, as shown by the copper test, added to the sugar under examination, in the form of corn glucose or starch-sugar. Figs. 39 and 39*a* show

Fig. 39.

the arrangement adopted. The middle portion of the saccharimeter is so modified as to admit of the interposition of a water-bath in the space ordinarily intended for the observation-tube alone. This is heated from below by two or four small spirit-lamps, and an opening is made in the cover of the water-bath for a thermometer whereby the temperature of the water is regulated. The form of the tube is shown in Fig. 39 *a*, which is merely the ordinary

Fig. 39 *a*.

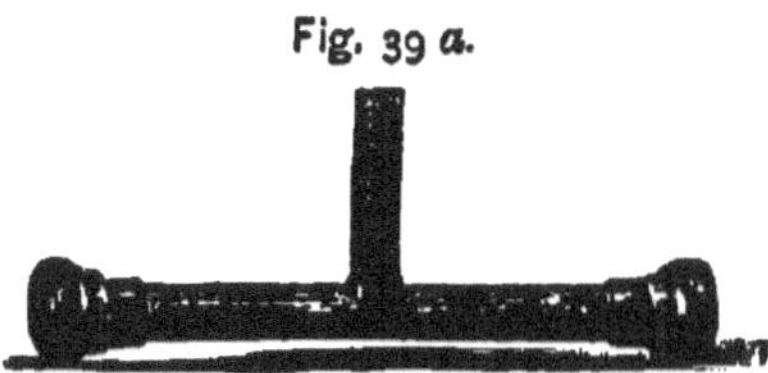

one provided with a tubule for the introduction of a thermometer into the tube itself.

This method in many cases is capable of giving useful results, and though a decided advance over previous methods for the optical estimation of sugar in the presence of starch-sugar, yet it must not be forgotten that when the composition of the adulterant varies considerably from the above standard, or that of any other standard taken, the results, considered quantitatively, will be misleading.

CHAPTER XV.

ESTIMATION OF MILK-SUGAR.

I. **By Fehling's Method.**—Milk-sugar reduces the alkaline solution of oxide of copper in a different proportion from dextrose or invert-sugar. One equivalent of milk-sugar reduces 7.40 to 7.67 eq. (Soxhlet *), 7.40 to 7.44 eq. (Rodewald and Tollens †). 10 c.c. of the standard copper liquor is equivalent to .067 gramme sugar.

$$\left.\begin{array}{r}\text{Copper} \times .7635 \\ \text{Copper oxide} \times .6096\end{array}\right\} = \text{milk-sugar.}$$

The estimation is precisely similar to that made for dextrose and invert-sugar, except that it is necessary to heat somewhat longer, as the reaction, though complete, does not take place so rapidly as with dextrose. Either the volumetric or gravimetric methods may be used.

To estimate the sugar in *milk*, it is necessary to coagulate the caseine with a few drops of hydrochloric or acetic acids, and filter, before proceeding with the operation.

II. **By the Optical Method.**—When the normal weight of 32.680 grammes for the Ventzke-Scheibler saccharimeter, and 20.50 grammes for the Soleil-Duboscq and other saccharimeters in which the normal weight is 16.19 grammes for cane-sugar, is taken, each degree of the scale, when the 200 mm. tube is used, corresponds to one per

* See references for Soxhlet's work, pages 201, 202.

† *Scheibler's Neue Zeit.*, iv. 67–86.

cent. milk-sugar. As milk-sugar exhibits the phenomenon of birotation, it is necessary to heat the freshly-prepared solution for a few minutes before taking the reading in the saccharimeter.

For the estimation of milk-sugar in *milk*, the fat and caseine must be first removed, the latter being strongly levo-rotatory; 50 c.c. of the milk is mixed in a porcelain dish with 25 c.c. lead solution of moderate strength, and the mixture heated to gentle ebullition and allowed to cool; it is then washed into a 100 c.c. flask, which is filled to the mark, and the solution filtered. The clear filtrate is then examined, the 200 mm. tube being used. The readings must be doubled on account of the dilution from 50 c.c. to 100 c.c. If the milk exhibits an acid reaction it must be neutralized with soda solution (Landolt).

CHAPTER XVI.

Estimation of Dextrose in Diabetic Urine.

I. BY THE OPTICAL METHOD.

For ordinary cases the mode of proceeding is exactly as in the case of dilute sugar solutions in chapters xi. and xii., the urine being decolorized with lead and bone-black when necessary. Owing to the fact that the specific rotatory power of dextrose is considerably lower than that of cane-sugar, when the various saccharimeters are employed the normal quantity to be weighed, in order that the readings may indicate percentages, must be greater, and in proportion to the relative specific rotations of the two sugars. Taking [66.5] D for cane-sugar, and [53] D for dextrose, we have 66.5 : 53 : : 26.05 : x = 32.683 grammes, which is the dextrose normal weight for the Soleil-Scheibler instrument. Calculating similarly for the others, we have, when the normal quantity for the saccharimeters is weighed and made to 100 c.c.,

1° of the Laurent and Soleil-Duboscq instruments = 2.031 grammes,
1° of the Soleil-Ventzke instrument = 3.268 grammes,
1° of the Wild instrument (sugar scale) = 1.255 grammes,
1° of the Mitscherlich instrument = 9.410 grammes,

in one litre.

Schmidt and Haensch have made a modification of the Soleil instrument, so that the scale reads directly the number of grammes dextrose in 100 c.c.; this is called the *diabetometer*.

When the urine contains albumen it must be removed before the sugar can be estimated, as the former body has a strong rotation to the left. For this purpose the secretion is heated in a dish, with acetic acid added to acid reaction, until the albumen separates in flocks, which is then filtered off, washed, and the urine with the washings made up to the initial volume; or the urine, acidified with acetic acid, may be diluted with a concentrated solution of sodium sulphate to double its bulk, when the albumen separates and may be filtered off.

Biliary acids, though right-rotating, are seldom present in quantities sufficient to affect the substantial accuracy of the optical method.

When the urine contains less than 2.00 grammes of sugar to the litre, or the normal secretion is to be tested, the above mode of proceeding is unsuitable. Landolt * gives the following method for use under these circumstances: To one or two litres of the urine neutral acetate of lead is added, and the solution filtered; the filtrate is mixed with basic lead acetate and ammonia, the precipitate formed containing all the sugar present. This precipitate is diffused in alcohol and treated with sulphuretted hydrogen gas, the lead sulphide filtered off, the solution decolorized, if necessary with animal charcoal, evaporated to a known volume, and tested in the saccharimeter. If biliary acids are present in the urine, they will be found in the alcoholic solution, and invalidate the optical test to some extent. To prove whether these acids are present or not, a portion of

---

* *Das optische Drehungsvermögen*, p. 185.

the alcoholic solution is evaporated to dryness, the residue taken up with water, and the solution obtained allowed to ferment with yeast for two days, or until the sugar is destroyed. If the filtered residual solution shows a right rotation, biliary acids are present, and a correction for them must be made.

II. BY FEHLING'S METHOD.

The Qualitative Test.—"Fifteen or twenty drops of the urine to be tested, previously decolorized with a little powdered bone-black and diluted with four or five c.c. of water, are treated with a half cubic centimetre of sodium or potassium hydrate solution, and then a very dilute solution of copper sulphate added drop by drop. Too large an amount of the copper salt should not be added, as in that case black oxide of copper separates on boiling, obscuring the red color of the cuprous oxide when only a small quantity of sugar is present. The clear blue solution is heated nearly to ebullition, without shaking, when a yellow cloud forms on the surface, followed by a precipitation of red cuprous oxide.

"A second mixture prepared in the same way is allowed to stand quietly, without previous heating, from six to twenty-four hours, when, if sugar is present, there will be a precipitate formed in this case also. This control experiment is of great importance, and ought never to be omitted, since most of the substances which reduce copper solution, like sugar, do so only when heated, or after prolonged boiling, and not, like diabetic sugar, in the cold" (Neubauer).

The Quantitative Estimation.—This determination is

made in all respects according to the ordinary volumetric method, or, for very accurate work, after the modification of Soxhlet. See chapter viii., section i. The gravimetric method is of doubtful accuracy, on account of the possible precipitation of earthy phosphates or other salts, under some conditions. It is best to decolorize the urine with a small quantity of powdered bone-black.

If albumen is present it must be separated by heating to boiling with a slight excess of acetic acid, filtering, and washing the precipitate.

Uric acid is probably the only body ordinarily present in urine which reduces the copper solution. According to many experiments of Neubauer, the uric acid in normal and diabetic urine has no appreciable effect on the results of the copper test. When uric acid is present to an abnormal amount, it may be removed by treating the solution, previously diluted to contain ½ per cent. sugar, with a slight excess of basic lead acetate, filtering, adding a solution of sulphurous acid until all lead is removed, and again filtering. The clear lead-free filtrate may be used for the sugar estimation.

CHAPTER XVII.

THE CHEMISTRY OF ANIMAL CHARCOAL.

Bone-Black—Bone-Char—Animal Char—Animal Black—Knochenkohle, Gr.—Charbon d'Os, Fr.

Composition.—Animal charcoal is the carbonaceous residue left by the distillation of bones in close vessels. Dr. Wallace gives as the average composition of a good char:

Carbon*	11.00
Carbonate of lime	8.00
Phosphates of lime and magnesia	80.00
Alkaline salts	.40
Sulphate of lime	.20
Oxide of iron	.10
Silica	.30
	100.00

* The carbon is much higher than that of the bone-black made in this country.

The following analyses give a good idea of the composition of American chars:

	1.	2.	3.	4.	5.	6.	7.	8.
Moisture	3.37	2.56	2.45	2.39	2.78	4.43	2.07	1.76
Carbon	8.05	7.67	7.76	7.55	8.17	8.70	8.47	9.08
Carb. lime	6.71	7.54	8.76	7.42	7.60	7.84	6.08	7.19
Sulphate of lime	trace.	.08	trace.	trace.	.06	trace.	trace.	trace.
Iron	.18	.30	.098	.17	.11	.04	.05	.10
Sand, clay, etc.	.43	.32	.32	.57	.57	....	....	....
Undetermined *	81.26	81.53	80.602	81.90	80.71	78.99	83.33	81.87
	100.00	100.00	100.00	100.00	100.00	100.00	100.00	100.00
Lbs. per cu. ft.	42.7	45.50	48.5	45.5	....	47.4	....	....

These analyses represent chars of the best quality, in grains of medium size.

**Alkaline salts*, phosphates of lime and MgO, etc.

SCHULTZ (*Stammer's Lehrbuch*, PAGE 528) GIVES ANOTHER SERIES:

	1.	2.	3.	4.	5.	6.	7.	8.	9.	10.	11.	12.
Carbon	12.46	8.57	9.34	8.83	8.49	7.94	7.41	7.20	7.11	6.08	6.07	5.73
Carbonate of lime	6.33	8.00	7.94	7.38	7.92	8.47	7.50	7.74	6.98	7.51	8.27	7.01
Sulphate of lime	.29	.45	.19	.11	.19	.20	.21	.15	.16	.20	.24	.17
Sulphide of calcium	.06	.07	....	.05	trace.	.03	.02	trace.	trace.	.04	.05	.06
Sand, etc	8.99	7.14	1.00	.48	.70	.91	2.00	.36	.81	5.60	3.07	.77
Soluble salts	1.29	1.14	.97	1.38	.99	.84	.89	1.23	1.17	.85	.77	1.00
Specific gravity	2.774	2.927	....	....	2.894	2.910	....	2.929	....	....	....	....
Weight of 1 litre in grammes	810	760	650	....	690	706	....	715	....	770	....	850
Decolorizing power in grains	14	21	....	45	53	....	....	54	....	55	....	....
" " in powder	13	15	....	....	14	....	....	14	....	12	....	....

Note.—Nos. 1 and 2 were large-grained; the rest in grains of medium size. All are from European sources.

Historical.—The decolorizing power of animal char was first noticed by Lowitz, but Figuier in 1811 proposed its use as a decolorizer. In 1812 Derosne introduced it into the sugar manufacture, and in 1821 Bussy and Payen thoroughly investigated its properties and mode of action. It was first used for sugar solutions on the large scale in a state of fine powder, and consequently after one operation it became useless for another. Dumont, however, in 1828, made a great advance in the practical application of animal black, by employing it in grains and filtering the sugar solution through a column of it. Afterwards, the char was submitted to a process of revivification by washing and burning, essentially the same as practised at the present time.

Mode of Action.—The effects of animal charcoal on sugar solutions may be classed under two heads, though the physical action is the same in either case—viz., *the removal of color*, and *the absorption of other soluble matters*. The two actions seem to be dependent to some extent, and the color of the filtered solution is in most cases a good index of the amount of purification effected, though not always, for the coloring matter in sugar solutions that have been much heated is entirely removed with great difficulty, while the absorption of salts and other matters takes place in a normal degree. Filhol gives a table showing the decolorizing powers of various substances as compared with bone-black washed with hydrochloric acid, which is called 100:

	Litmus.		Red Wine.		Molasses.	
	Cold.	Hot.	Cold.	Hot.	Cold.	Hot.
Hydrated sesquioxide of iron..	128.9	96.86	54.54	72.72	51.91	52.24
" oxide of lead........	....	....	....	79.41	103.83	84.37
Barium sulphate..............	....	....	....	25.96	46.15	25.96
Calcic phosphate or carbonate..	109	90.28	....	42.18	49.13	42.18
Magnesium carbonate........	....	....	77	....	....	....
Bone-black..................	100	100	100	100	100	100

The absorptive power of bone-black is owing to the presence of carbon in a minute state of division. The phosphate of calcium constitutes a framework, as it were, for the carbon, and, after the calcination of the bones, remains in a very porous condition; hence the lighter the char for a given bulk the better it absorbs.

Presence of Nitrogen.—The carbon contains from one to one and a half per cent. of nitrogen, which diminishes to about one-half per cent. when the char has been used some time. This substance seems necessary for the decolorizing effect, as no vegetable charcoal destitute of nitrogen has the same properties. Nitrogenous chars prepared in different ways have the property of absorbing color in various degrees. A table from Muspratt illustrates this:

	Decol. Power.
Ordinary animal black..............................	1.
" " treated with hydrochloric acid...	1.6
Ordinary black calcined with K_2CO_3.................	20.0
Blood " "	20.0
" " chalk..................	11.0
" " phosphate of lime.......	10.0
Albumen " K_2CO_3................	15.5
Gluten " "	15.5
Oil " phosphate of lime.......	1.9

Absorbing Power of Char.—Brimmeyr's experiments,

confirmed by Schultz, on the absorbing power of animal charcoal, gave rise to the following conclusions: 1. The absorptive power does not depend on the mechanical structure, but upon the amount of carbon contained. 2. Char which has lost its power for absorbing one substance is capable of taking up another body of a different chemical nature. 3. The quantities of matter absorbed by bone-char of various kinds are, when considered in relation to the amount of carbon present, really equivalent, and probably independent of the varying chemical nature of the absorbed substance. 4. Bone-char acts the quicker and better the less its capillary structure has been altered by mechanical or chemical means.

The following analyses, taken from actual work in a sugar refinery, show the absorptive action of char for soluble impurities:

	Raw Liquor.	Filtered Liquor.	Char Washings.
Sugar	93.50	95.30	78.50
Grape-sugar	2.14	2.25	3.23
Organic matter not sugar	3.56	2.00*	11.05
Ash	.80	.45†	7.22
	100.00	100.00	100.00

* 43.82 per cent. absorbed. † 43.75 per cent. absorbed.

Walkoff * gives an admirably clear, graphical representation of the progress of a filtration, showing the absorption of alkalies and coloring matter, and the progressive purification of the sugar solution (Fig. 40). The perpendicular

* *Traité Complet*, tome ii. 191.

lines show the hours of filtration, and the others the relative proportion of sugar in dry substance, alkalinity, and decolorization during the progress of the operation.

Absorption of Salts and Organic Matters.—The soluble substances taken up by char are either organic or mineral, the former consisting of gums, coloring matter, albumen, etc., and the latter of inorganic bases combined with organic or mineral acids. The organic bodies, notably albumen, are retained by the char with great tenacity, so that long washing with hot water, or even steaming,

Fig. 40.

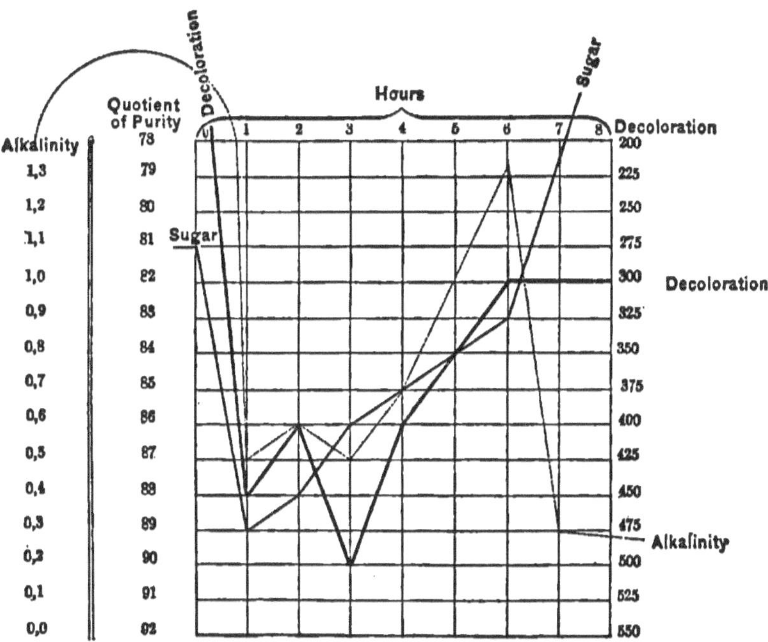

fails to remove all of the absorbed material; some inorganic salts are also obstinately retained. The soluble matters submitted to the action of the char are taken up in varying amount, depending on the nature of the body.

Walkoff,* working with weak solutions of potash and soda salts, arrives at the following results, the conditions being the same in all experiments, and the temperature 15° C.:

	Per cent. absorbed.		Per cent. absorbed.
Potassium hydrate (at 60° C.)..	13.5	Sodium carbonate...........	24.
" " (at 15° C.)..	16.6	" " (at 60° C.)..	18.3
" carbonate.........	25.0	" phosphate...........	32.3
" phosphate.........	30.7	" "	28.0
" nitrate............	6.5	" nitrate..............	5.0
" chloride..........	3.0	" sulphate............	20.4
" "	1.3	Magnesium sulphate.........	49.0
" citrate............	12.2	Sodium chloride.............	1.
" sulphate.........	22.4		

Bodenbender† has also examined the subject and extended the research so as to include salts of organic acids. In the table, under I. are included dilute solutions of the salts, with 5 per cent. of cane-sugar added, and under II. more concentrated solutions without sugar:

	Per cent. absorbed.	
	I.	II.
Sodium acetate......................	....	21.50
Potassium acetate	28.70	16.50
" oxalate....................	....	48.10
Sodium "	....	69.97
Potassium citrate..................	....	45.40
Sodium "	....	48.20
" chloride....................	13.51	8.10
Potassium chloride.................	11.75	9.15

The absorbing powers of char for different alkaline salts were found to be in the following order, commencing with the weakest: Potassium chloride, sodium chloride, potas-

* *Traité Complet*, tome ii. 206.

† *Stammer's Jahresb.*, x. 239.

sium nitrate, sodium nitrate, potassium acetate, sodium acetate, potassium sulphate, sodium sulphate, magnesium sulphate, potassium carbonate, sodium carbonate, and sodium phosphate.

The amount of purification effected by char in a sugar solution *is directly as the amount of the former used to a given weight of sugar, as the temperature, and as the time which the solution is in contact with the coal.*

Marks of a Good Char.—Good animal charcoal has a dull black color, without presenting any appearance of incipient fusion on the surface, and does not contain an undue proportion of cellular particles, which come from small and inferior bones, and have by no means the decolorizing effect of the char made from the large bones. It should adhere strongly to the tongue, and not contain much fine powder, but be hard and tough to resist the great wear to which it is subjected during filtration and revivification. The size is regulated by the density and temperature of the liquor to be filtered.

Revivification.—After bone-char has served the purposes of filtration, water as hot as possible is run in at the top of the filter, which displaces in part the sugar solution remaining in contact with the char, and at the same time it mixes with the rest, forming a dilute solution of sugar and the impurities taken up from the liquor ; this dilute solution is known as "sweet water." By the action of the heated water the char gives up the greater portion of the absorbed matter, which goes in part to the sweet water and the remainder to the "*waste water*," which is the wash-water that no longer contains sufficient sugar to make it profitable to save. The "sweet water" is generally boiled down with one of the lower products, and should on no account be used to dissolve comparatively pure raw sugars for refining.

From the above it follows that the purification of sugar solutions by bone-black consists in removing the impurities from the first products, but, instead of eliminating them entirely, adding a large portion of them to the lower or half-refined products, where their injurious influence comes less into play.

The second step in the revivification consists in heating the washed char in closed retorts, out of contact with the air, at a sufficiently high temperature to perfectly carbonize any organic matter remaining in it, and to bring the char back to the physical condition in which its absorbing properties are exerted to their fullest extent.

Alteration by Use.—The following analyses, taken from actual work, show the progressive changes that have taken place in bone-black used in a refinery where raw sugars from the cane were worked:

1877.	Average of new black.	April 4.	May 1.	May 22.	June 3.	July 12.	Aug. 7.	Sept. 20.	Oct. 15.	Nov. 6.	Dec. 1.	Dec. 8.	Jan. 8, 1878.
Moisture	3.37	.68	....	....	....	....	....	....	....	....	....	....	....
Carbon	8.05	7.65	8.05	8.38	8.74	9.62	9.15	9.44	9.59	9.89	10.07	10.07	10.01
Carbonate of lime	6.71	5.03	5.47	6.11	5.40	4.86	4.34	4.21	4.33	4.12	4.08	4.16	4.14
Iron	.18	.21	.21	.23	.19	.24	.20	.25	.32	.34	.32	.33	.36
Insoluble matter	.43	.45	.46	.58	.26	.48	.44	.38	.34	.44	.33	.63	.42
Sulphate of lime	....	....	.35	....	....	.49	.34	.44	....	.49	.31	....	....
Sulphide of calcium	....	....	....	....	....	....	....	....	....	....	....	....	.41
Lbs. per cu. ft.	42.7	49.9	51.3	52.9	52.9	53.9	51.9	56.4	55.5	55.3	58.8	61.7	58.7
Decolorizing power—per cent. of color absorbed from a sugar solution	58.2	....	....	54.	52.7	....	52.3	51.8	51.8	43.5	47.6	48.0	....

An examination of the table shows that the char by use undergoes a change, which will be examined *seriatim* in relation to the various constituents of the black.

Carbon increases owing to the fact that some organic matters absorbed are held with great tenacity, and no practical amount of washing is sufficient to remove all traces of these bodies. The consequence is that, in the burning in the process of revivification, the residual organic matter is carbonized in the pores of the coal, and, instead of increasing the decolorizing effect of the coal, the opposite is the case, as a quantity of inert non-nitrogenous carbon is deposited in the body of the grain, making it more dense and decreasing the amount of cellular surface. A large percentage of carbon in a new char is often a mark of poor quality, being caused by imperfect and insufficient burning of the bones. As a rule old and new char may be distinguished by the proportion of carbon contained.

Carbonate of Lime.—The tenor of this salt is rapidly reduced when the char is first used for filtration, and then generally remains stationary for a long time at from 4.50 per cent. to 3.50 per cent., but finally, in very old chars, the percentage may be lowered to less than the last figure ; in this case there is too little of the salt for normal working. Its office is mainly to ensure neutral liquors by saturating any free acid which may exist in the solution, or may be formed by the lactic-acid fermentation, where the temperature of filtration is too low. As the solutions ought always, whether in filtering or washing, to be not lower than 180° F., there can be no danger of this fermentation so long as this condition is complied with. By taking proper care to have the raw liquor sufficiently limed, and the temperature high enough in the filters, the amount of carbonate of lime in the char will remain high enough almost indefinitely.

In the beet-sugar manufacture, where the liquors and juices often contain a large excess of caustic and other lime-salts, the percentage of carbonate is apt to be high rather than low. It has in that case to be removed by treatment with hydrochloric acid. An excess of the carbonate is injurious to the char in the same way as is any other insoluble inert body, which stops up the pores and reduces the available filtering surface. Hard water containing much carbonate of lime is not suitable for washing char, as the lime-salt is retained.

Alkaline Salts.—These consist largely of ammoniacal compounds, and are only found in considerable quantity in new char. If suffered to remain they go into the liquor and act injuriously by their melassigenic properties. On this account new char should be thoroughly washed and burned before using.

Sulphate of Lime.—This salt acts by filling up the pores of the coal, and may be derived from the sugar treated or the water used in washing the char; it is strongly retained by the char, but thorough washing is the remedy in this as in many other cases.

Iron is a highly injurious body, and is derived from the sugar treated or from rusted, insufficiently-painted filters and piping, and also from the retorts of the kilns in which the black is burned. When existing in the char it is sure to get into the filtered liquors, and especially the sweet waters, more particularly when they are a little acid. It accumulates in the yellow sugars or lower products of the refiner, giving them an undesirable dull grayish cast which greatly lowers their marketable value. Such sugars also darken tea to which they are added, by the formation of tannate of iron. All new black purchased for refining purposes should carry very low percentages of iron.

To prevent iron from getting into the char in the course of manufacture, the filters should be well scraped and painted as often as is necessary, and the liquors should be neutral.

Insoluble matter which resists the solvent action of strong acids on the char consists partly of quartz, sand, or clay—which in moderate amount is not objectionable—and also of hydrated silica derived from weak sugar solutions that have soured and precipitated the dissolved silica. This is caught in the char, and acts in the manner of sulphate of lime or other finely-divided matters.

Sulphide of calcium is apt to accumulate in char that has been used some time and which contains much sulphate of lime. The sulphate is reduced in contact with the organic matter or carbon in the reburning of the char, forming the soluble sulphide which goes into the liquors. Sulphide of calcium, coming in contact with the iron in the liquors, strikes a yellowish green color, which develops, even after the solution has run off the coal, from the formation of ferrous sulphide, and very seriously interferes with the operation of making salable sugars, especially for the lower products. Calcium sulphide is one of the worst impurities that can exist in bone-black used for purposes of filtration, and any sample containing more than a very small quantity will fail to give a satisfactory working on the large scale. In contact with an acid sugar solution sulphide of calcium also gives off hydrosulphuric acid, which at favorable temperatures is believed to predispose the sugar solutions to fermentation. In the series of analyses given on page 305, the last, showing .41 per cent. of calcium sulphide, represented char which was rejected as being no longer fit for filtration, and chiefly owing to presence of the sulphide.

Nitrogen appears to greatly aid in the decolorizing action of the carbon in animal black, and, as a rule, the higher the amount the better the char.

It is often a question submitted to the chemist as to whether a given char is so far exhausted as to decolorizing properties that it would be desirable to replace it with new. Many things have to be considered in this relation, the chemical analysis alone not always being a sufficient guide, as chars which analyze poorly sometimes decolorize very well. No general rule can be laid down for the matter, and the chemist will have to rely largely upon the results obtained in working on the large scale with the black in question, extending over a sufficient length of time, and including all the necessary analytical details; above all, reliance should be placed on his general experience gained in this special department.

There is appended a series of analyses of bone-blacks, I.* showing both old and new chars, and II.† char of English and American origin that has been used in filtration:

---

* Maumenè, *Traitè Complet.*

† W. Arnott, *Amer. Chemist*, i. 216.

I.

Carbon........	10.21	8.44	9.88	12.07	10.65	24.22	26.12
Phosphate of lime..	76.94	80.31	80.11	76.35	78.52	64.35	62.40
Carbonate of lime.	7.42	8.77	7.76	7.09	7.21	4.82	3.35
Sulphate of lime...	.12	.40	.08	.11	.20	.84	.88
Sulphide of calcium	.01	.04	....	....	.08	.14	.16
Alkaline salts.....	.67	.35	.04	.33	.17	.12	.24
Oxide of iron.....	.22	.43	.17	.12	.06	.43	.56
Sand..............	.34	1.71	.92	.79	.73	1.35	4.08
Water............	4.07	.55	1.04	1.14	2.38	3.73	2.21
	100.00	100.00	100.00	100.00	100.00	100.00	100.00
Real density......	2.89	2.928	2.903	2.937	2.943	2.939	2.937
Apparent density:							
In powder......	1.070	.996	.944	1.081	.975	1.144	1.388
In grain........	.776	.804	.769	.778	.771	1.088	1.196
Decolorizing power:							
In powder......	142	94	116	104	165	62	51
In grain.........	93	71	88	65	102	8	7

II.

Origin.	Greenock.	Greenock.	Liverpool.	Liverpool.	London.	London.	New York.	New York.
Carbon..............	12.90	16.35	9.02	8.24	11.40	11.20	10.20	10.45
Phosphates.........	82.04	77.93	85.44	87.48	80.61	83.80	83.43	84.50
Carbonates........	3.23	3.30	3.10	2.00	5.98	3.33	4.76	3.75
Sulphates..........	.27	.29	.42	.58	.92	.14	.17	.27
Oxide of iron......	.51	.33	.48	.57	.43	.32	.51	.43
Alkaline salts......	.25	.20	.20	.15	.20	.20	.10	.15
Sand, etc..........	.80	1.60	1.34	.98	.46	1.01	.83	.45

CHAPTER XVIII.

The Analysis of Animal Charcoal.

ESTIMATION OF WATER.

Dry for two hours at 140° C. The sample should not be powdered.

ESTIMATION OF CARBON.

Dissolve four or five grammes of the finely-powdered char in about 35 c.c. of pure hydrochloric acid diluted with its bulk of distilled water; heat on a water-bath in a flask or beaker-glass for half an hour, with frequent agitation, until the soluble part has all been taken up. Dilute to about 200 c.c. with hot distilled water, allow to settle, and pour on a filter that has been previously washed with dilute acid dried at 100° and weighed. When the liquid has all filtered, add more hot water to the flask, shake well, allow to subside, and pour off the clear solution from the undissolved matter; add water to the flask a third time, with a little hydrochloric acid, and transfer the carbon to the filter in the usual way. Continue the washing on the filter with hot water, at first acidulated, and finally with pure water, until the washings have no longer an acid reaction, or a drop, when evaporated on platinum foil, leaves little or no residue. Dry the filter at 100° until it

Fig. 41.

ceases to lose weight, the weighing being performed in the same vessel (watch-glasses or a weighing-flask, Fig. 41) in which the filter had been previously tared. After the last weighing, transfer the filter with the carbon to a weighed crucible, burn off the carbon, and reweigh. The residue in the crucible, after the subtraction of the filter-ash, constitutes the *insoluble residue*, which, taken from the last weight at 100°, gives the amount of pure carbon.

Example:

Amount of char taken....................	4.500	grammes.
Residue + weighing-flask + filter at 100°.	20.570	"
Weighing-flask + filter at 100°...........	20.150	"
Carbon + insoluble matter................	.420	"
Insoluble matter..........................	.019	"
Carbon....................................	.401	"
Crucible after ignition...................	15.140	"
"	15.120	"
	.020	"
Filter-ash................................	.001	"
Insoluble matter..........................	.019	"

$$\frac{.401 \times 100}{4.50} = 8.91 \text{ per cent. carbon.}$$

$$\frac{.019 \times 100}{4.50} = .42 \text{ per cent. insoluble matter.}$$

Note.—Char should be weighed in a close vessel for purposes of analysis, either between watch-glasses or in a weighing-flask, as, according to the state of the atmosphere, or the amount of water in the char itself, there will be a gain or loss during the time necessary to take an accurate weight.

Fig. 42.*

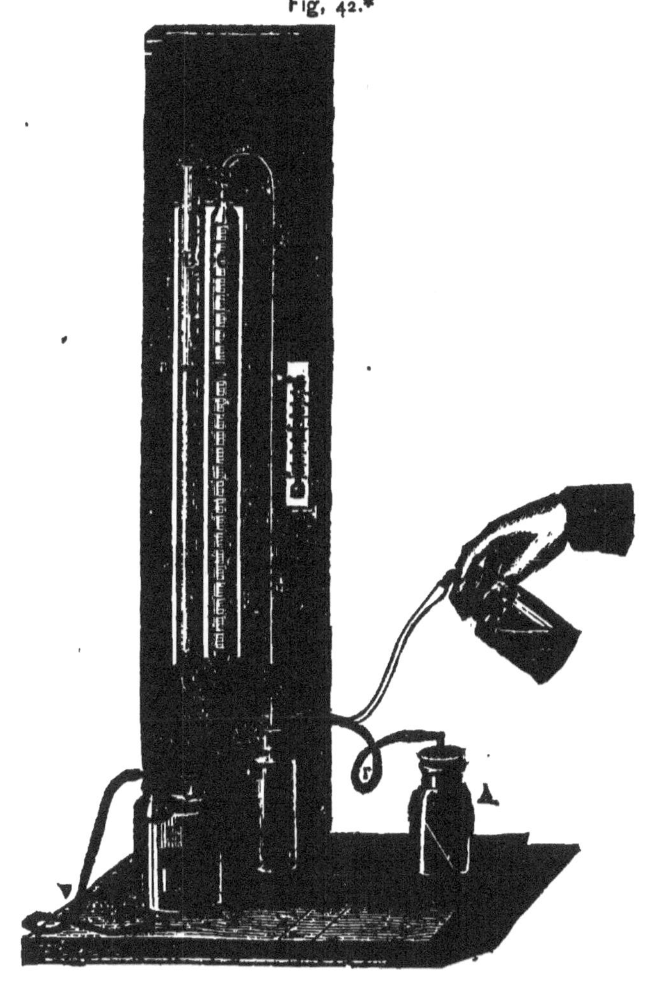

ESTIMATION OF CARBONATE OF LIME.

For work where ordinary accuracy is required, this estimation had best be performed according to Scheibler's

* The author is indebted to Messrs. Elmore & Richards, of New York, for the above engraving.

process.* The results for low percentages are accurate enough for all technical purposes. We give Scheibler's description. The apparatus is represented by Fig. 42, and consists of the following parts:

1. The evolution-flask A, in which the assay is acted upon by hydrochloric acid, which is placed in the rubber tube S. The glass stopper of A is perforated, and carries a tube, to which is joined a rubber tube, *r*, connecting A with B. The latter has a gum stopper fitted with three glass tubes; the one joined to *r* extends a short distance into the vessel, and has fastened to it, by the neck, a thin caoutchouc bag, K, capable of being easily distended by a slight pressure; *q* is closed by a pinchcock while the estimation is being made, and serves to bring B into communication with the air when necessary. The glass tube *u* also passes through the stopper of B and connects with

(2) The graduated tube C, which is divided into twenty-five equal parts (about 4 c.c. each), each division being subdivided into tenths. The lower end of this is in communication with

(3) The straight control-tube D, open at the upper end, and at the lower having a tube of smaller calibre passing to the bottom of the two-necked flask E, as shown in the figure, the connection between the two being regulated by the pinchcock *p*. E is the reservoir for the water, and C and D are filled from it by pressure exerted by the breath of the operator through *v*, the cock *p* preventing the reflux of the water. C, D, and *u* are fastened to an upright board by suitable means, and the bottles are supported on a shelf fastened to the upright board.

* *Stammer's Jahresb.*, 1861–2, 244.

In addition to the apparatus, the following requisites are necessary to the performance of the test:

1. A normal weight of 1.702 grammes.
2. A centigrade thermometer graduated from 12° to 30°.
3. Diluted hydrochloric acid of specific gravity 1.120.
4. A solution of chloride of copper.
5. A solution of ammonium carbonate.

For the execution of a test the normal quantity of pulverized char (1.702 grammes) is placed in A, which must be dry, and the tube S, filled with acid to the mark, is carefully placed in the bottle. E is then filled with water, and the operator forces the liquid into D and C until it reaches a little above the zero-point in C, when it is allowed to flow out by opening p until the level in C is at 0. Care must be taken that the water is not caused to overflow into B, for in that case the apparatus would have to be taken apart and dried. The stopper being now placed in A, a connection with B is made by the tube r. If the level of liquid in D and C are then unequal, the equality may be restored by opening the cock q for a few seconds, and which for the rest of the operation remains closed.

The test may now be proceeded with. The vessel A is held, as shown in the cut, so that the acid may come in contact with the char, and the bottle gently shaken to cause the acid to thoroughly mix with the assay. The pressure of the gas evolved distends the rubber bag and depresses the column of water in C. The cock p is now opened to let the water in D flow out, the operator aiming to keep the level in C and D as near the same as possible during the progress of the determination. When all the gas has been given off, and the level of the liquid in C becomes stationary, p is closed after bringing the water in D to the same level as that in C, and the volume and temperature read off.

New char sometimes contains a small quantity of caustic lime. When this is the case the finely-powdered char before being tested is evaporated on the water-bath, after being thoroughly moistened with a solution of carbonate of ammonia.

The presence of sulphide of calcium in char introduces a slight error in this method, as it is decomposed in contact with the acid, setting free sulphuretted hydrogen, which would be reckoned as carbonic acid gas. This difficulty may be met by the addition of a small quantity of cupric chloride to the acid used.

The apparatus should be placed in a position where the temperature is as equable as possible. The estimation is made in duplicate, and the average taken as the true result. The following table gives the percentage of carbonate of lime from the volume and temperature readings:

TABLE FOR CALCULATING THE PERCENTAGE OF CARBONATE OF LIME FROM THE VOLUME OF CARBONIC ACID: FOR USE WITH SCHEIBLER'S CALCIMETER.

Volume read.	DEGREE CENTIGRADE.																		
	12°	13°	14°	15°	16°	17°	18°	19°	20°	21°	22°	23°	24°	25°	26°	27°	28°	29°	30°
1	.80	.80	.79	.79	.79	.78	.78	.77	.77	.77	.76	.76	.76	.75	.75	.74	.74	.73	.73
2	1.88	1.87	1.86	1.86	1.85	1.84	1.83	1.82	1.81	1.80	1.79	1.79	1.78	1.77	1.76	1.75	1.74	1.73	1.72
3	2.95	2.94	2.92	2.91	2.90	2.89	2.87	2.86	2.85	2.83	2.82	2.80	2.79	2.77	2.76	2.74	2.73	2.72	2.71
4	4.01	4.00	3.98	3.96	3.94	3.93	3.91	3.89	3.87	3.85	3.83	3.81	3.79	3.77	3.75	3.73	3.71	3.70	3.68
5	5.07	5.05	5.03	5.00	4.98	4.96	4.93	4.91	4.87	4.86	4.84	4.81	4.79	4.76	4.74	4.71	4.69	4.67	4.65
6	6.11	6.09	6.06	6.03	6.01	5.98	5.95	5.92	5.89	5.86	5.83	5.81	5.78	5.75	5.71	5.68	5.65	5.63	5.61
7	7.14	7.12	7.09	7.06	7.02	6.99	6.96	6.92	6.89	6.86	6.82	6.79	6.75	6.72	6.68	6.65	6.61	6.58	6.56
8	8.17	8.14	8.11	8.07	8.03	8.00	7.96	7.92	7.88	7.84	7.80	7.76	7.72	7.68	7.64	7.59	7.55	7.53	7.49
9	9.19	9.16	9.12	9.07	9.03	8.99	8.95	8.90	8.86	8.82	8.77	8.73	8.68	8.64	8.59	8.55	8.50	8.46	8.42
10	10.20	10.16	10.12	10.07	10.02	9.98	9.93	9.88	9.83	9.79	9.73	9.68	9.63	9.58	9.53	9.48	9.43	9.39	9.34
11	11.20	11.15	11.10	11.05	11.00	10.95	10.89	10.84	10.79	10.74	10.68	10.63	10.57	10.52	10.46	10.41	10.35	10.30	10.25
12	12.20	12.15	12.09	12.03	11.98	11.92	11.87	11.81	11.75	11.69	11.64	11.58	11.52	11.46	11.40	11.33	11.27	11.22	11.16
13	13.20	13.14	13.08	13.02	12.96	12.90	12.84	12.78	12.72	12.65	12.59	12.53	12.46	12.40	12.33	12.26	12.20	12.14	12.07
14	14.20	14.14	14.07	14.01	13.94	13.88	13.81	13.75	13.68	13.61	13.54	13.48	13.41	13.34	13.26	13.19	13.12	13.05	12.99
15	15.20	15.13	15.06	14.99	14.92	14.85	14.78	14.71	14.64	14.57	14.50	14.42	14.35	14.27	14.20	14.12	14.04	13.97	13.90
16	16.20	16.13	16.05	15.98	15.91	15.83	15.76	15.68	15.61	15.53	15.45	15.37	15.29	15.21	15.13	15.05	14.97	14.89	14.81
17	17.20	17.12	17.04	16.97	16.89	16.81	16.73	16.66	16.59	16.49	16.41	16.32	16.24	16.15	16.07	15.98	15.89	15.81	15.72
18	18.20	18.12	18.03	17.95	17.87	17.79	17.70	17.62	17.53	17.45	17.36	17.27	17.18	17.09	17.00	16.91	16.82	16.73	16.63
19	19.20	19.11	19.03	18.94	18.85	18.75	18.67	18.59	18.50	18.40	18.31	18.22	18.13	18.03	17.94	17.84	17.74	17.64	17.55
20	20.20	20.11	20.02	19.93	19.83	19.74	19.65	19.55	19.46	19.35	19.27	19.17	19.07	18.97	18.87	18.77	18.66	18.56	18.46
21	21.20	21.10	21.01	20.91	20.81	20.72	20.62	20.52	20.42	20.32	20.22	20.12	20.01	19.91	19.80	19.70	19.59	19.48	19.37
22	22.20	22.10	22.00	21.90	21.80	21.70	21.59	21.49	21.39	21.28	21.17	21.07	20.96	20.85	20.74	20.63	20.51	20.40	20.28
23	23.20	23.09	22.99	22.88	22.78	22.67	22.56	22.46	22.35	22.24	22.13	22.02	21.90	21.79	21.67	21.55	21.44	21.31	21.20
24	24.20	24.09	23.98	23.87	23.76	23.65	23.54	23.43	23.31	23.20	23.08	22.97	22.85	22.73	22.61	22.48	22.36	22.23	22.11
25	25.20	25.08	24.97	24.86	24.74	24.63	24.51	24.39	24.28	24.16	24.04	23.91	23.79	23.67	23.54	23.41	23.28	23.15	23.02

An example will show the use of the table: A sample of char gave a volume of 15.4 at 25° C.

15.0 volumes = 14.27 at 25°
.4 " = .37 at 25°

14.64 per cent. calcium carbonate.

The carbonate of lime in animal black may be estimated very accurately by several processes depending upon the expulsion of the gas, determining the weight lost, and calculating the carbonate of lime from the carbonic acid lost: $CO_2 \times 2.2727$ = carbonate of lime. For details of these and other methods the reader must be referred to standard works on analytical chemistry.

Calculation for Removal of Carbonate by Acid.—In the beet-sugar manufacture, and in refining where the water used for washing the char is very hard, calcic carbonate accumulates in the char to an abnormal extent, and it is often desirable to remove the excess by washing with hydrochloric acid. Taking 7 per cent. as the normal amount, Scheibler has given a table whereby the amount of hydrochloric acid of any strength required to reduce the carbonate to the prescribed limit may be calculated:

Degree Baume.	Sp. Gr. at 15° C.	Percentage of Hydro-chloric Ac.	Percentage of $CaCO_3$ dissolved by Acid.	Quantity of Hydrochloric Acid necessary for the Solution of 1, 2, . 9 parts of Carbonate of Lime.								
				1	2	3	4	5	6	7	8	9
25.0	1.210	42.4	58.088	1.7217	3.4434	5.1651	6.8868	8.6085	10.3302	12.0519	13.7736	15.4953
24.5	1.205	41.2	56.444	1.7718	3.5437	5.5155	7.0874	8.8592	10.6310	12.4029	14.1747	15.9466
24.0	1.199	39.8	54.526	1.8342	3.6683	5.5025	7.3367	9.1709	11.0050	12.8392	14.6734	16.5075
23.5	1.195	39.0	53.430	1.8718	3.7436	5.6154	7.4872	9.3590	11.2308	13.1026	15.0744	16.8462
23.0	1.190	37.9	51.923	1.9261	3.8522	5.7784	7.7045	9.6306	11.5567	13.4828	15.4090	17.3351
22.5	1.185	36.8	50.416	1.9837	3.9674	5.9511	7.9348	9.9185	11.9022	13.8859	15.8696	17.8533
22.0	1.180	35.7	48.909	2.0448	4.0896	6.1344	8.1792	10.2240	12.2688	14.3136	16.3584	18.4032
21.5	1.175	34.7	47.539	2.1037	4.2075	6.3112	8.4150	10.5187	12.6224	14.7262	16.8299	18.9337
21.0	1.171	33.9	46.443	2.1534	4.3068	6.4602	8.6136	10.7670	12.9204	15.0738	17.2272	19.3806
20.5	1.166	33.0	45.210	2.2121	4.4242	6.6363	8.8484	11.0605	13.2726	15.4847	17.6968	19.9089
20.0	1.161	32.0	43.840	2.2813	4.5625	6.8438	9.1250	11.4063	13.6875	15.9688	18.2500	20.5313
19.5	1.157	31.2	42.744	2.3397	4.6795	7.0192	9.3590	11.6987	14.0384	16.3782	18.7179	21.0577
19.0	1.152	30.2	41.374	2.4172	4.8344	7.2517	9.6689	12.0861	14.5033	16.9205	19.3378	21.7550
18.0	1.143	28.4	38.908	2.5704	5.1408	7.7113	10.2817	12.8521	15.4225	17.9929	20.5634	23.1338
17.0	1.134	26.6	36.442	2.7444	5.4887	8.2331	10.9774	13.7218	16.4662	19.2105	21.9549	24.6992
16.0	1.125	24.8	33.976	2.9436	5.8871	8.8307	11.7742	14.7178	17.6613	20.6049	23.5484	26.4920
15.0	1.116	23.1	31.647	3.1602	6.3206	9.4805	12.6407	15.8009	18.9610	22.1212	25.2814	28.4415
14.0	1.108	21.5	29.455	3.3954	6.7907	10.1861	13.5814	16.9768	20.3721	23.7675	27.1628	30.5582
13.0	1.100	19.9	27.263	3.6683	7.3367	11.0050	14.6734	18.3417	22.0100	25.6785	29.3467	33.0151

Example: A char contains 12.30 per cent. calcic carbonate, and the acid at command has a density of 1.166, or 20° B. Now, 12.30 — 7.00 = 5.30 per cent. carbonate to be removed. From the table we find—

5.0 parts	$CaCO_3$	require	11.40 parts acid,	
.3 "	"	"	.68 "	
			12.08 parts, of sp. gr. 1.166,	

or, in a ton of 2000 lbs. of char,

2000 × 12.08 per cent. = 241 lbs. of commercial acid of the indicated strength.

ESTIMATION OF CALCIC SULPHATE.

For this and the succeeding determination the char should be very finely pulverized and passed through an 80-mesh sieve. Twenty grammes are taken, placed in a porcelain dish on a water-bath, moistened with distilled water, 80 c.c. of pure concentrated hydrochloric acid added, and the whole heated for an hour with frequent stirring. At the end of that time the semi-fluid mass is washed into a 250-c.c. flask, diluted to the mark, and the mixture filtered. To 200 c.c. of the clear filtrate, corresponding to 16 grammes of the original substance, is added, in the heat, its bulk of water, together with a slight excess of barium chloride, and allowed to stand at rest from six to twelve hours. The precipitated barium sulphate is now filtered from the clear solution, and, after washing two or three times with boiling water in the beaker, is treated with about 5 c.c. of a strongly acid solution of ammonium acetate, heated for five minutes, diluted with boiling water, and the precipitate and fluid transferred to the filter. After a further washing, the filter is dried and the weight of the barium sulphate determined in the usual manner.

Barium sulphate × .58324 = calcium sulphate.*

ESTIMATION OF CALCIUM SULPHIDE.

Twenty grammes of the finely-powdered char are treated in a porcelain dish on a water-bath, after first moistening with water, with 40 c.c. of fuming nitric acid free from sulphuric acid, added in small portions at a time to prevent too violent a reaction. The mixture is heated, with frequent stirring, for half an hour, when 40 c.c. of pure concentrated hydrochloric acid are added gradually, and the whole kept heated, with stirring as before, for twenty minutes longer. The contents of the dish are now transferred to a 250-c.c. flask, and when cold the fluid is diluted to the mark and filtered; 200 c.c. of the filtrate, corresponding to 16 grammes of char, after dilution with an equal volume of water, are treated with a slight excess of barium chloride, and the amount of sulphate formed determined as in the estimation of calcium sulphate.

It is of the first importance that, in this and the preceding estimation, *the reagents used should be absolutely free from sulphur in any form.*

For the calculation of the results, the amount of the barium salt found in the determination of calcic sulphate is subtracted from that as obtained above, and the remainder

* The error owing to the volume occupied by the undissolved carbon has been experimentally proved to be without sensible effect on the results, and likewise that from the slight solubility of barium sulphate in a liquid containing a considerable amount of free hydrochloric or nitric acids.

is the barium sulphate corresponding to the calcic sulphide:

Barium sulphate × .3089 = calcic sulphide.

Example: 10 grammes of char containing .50 per cent. sulphate of lime, when treated for the estimation of sulphide, gives .230 gramme $BaSO_4$. Now, the barium sulphate from the calcic sulphate would be

$$10.000 \times .0050 = .050 \text{ gramme } CaSO_4, \text{ and}$$

$$\frac{.050}{.58324} = .0857 \text{ gramme } BaSO_4.$$

$$.2300 - .0857 = .1443 \text{ gramme } BaSO_4,$$

furnished by the oxidation of the sulphide; hence,

$$\left(\frac{.1443 \times .3089}{10}\right) \times 100 = .445 \text{ per cent. } CaS.$$

Iles and Fahlberg's * method, though more tedious in execution than the above, gives very good results

ESTIMATION OF CALCIC PHOSPHATE.

About one gramme of the powdered char is ignited in a crucible until the carbon is burned off; the residue is then dissolved in 50 c.c. pure nitric acid, and the solution made up with water to 100 c.c.; 25 c.c. of this solution is taken and treated gravimetrically by precipitation with molybdenum solution, or by the volumetric method with uranium acetate. For details of these methods the reader is referred to Fresenius's or other standard works on analytical chemistry.

---

* *Ber. Chem. Gesell.*, 1879, xi. 1187.

This determination is rarely necessary, except upon the exhausted char, when it is desired to estimate its value for fertilizing purposes.

ESTIMATION OF THE IRON.

The iron may be determined in the filtrate from the carbon, or from an ignited portion of the char, dissolved in strong hydrochloric acid. When the tenor of the iron is very low about 10 grammes should be taken for the assay. To the strongly acid solution, platinum foil and a piece of iron-free zinc are added to reduce the sesquioxide to protoxide. When the liquid no longer gives a red coloration with a drop of ammonic sulphocyanate solution, the reduction is complete. The iron in the ferrous condition is then determined by a standard solution of potassium permanganate.

Preparation of the Standard Solution.—This solution is prepared by dissolving about 2½ grammes of the crystallized salt in water and diluting to one litre. To find the exact amount of iron that the solution is equivalent to, 1.500 grammes pure crystallized oxalate of ammonia are dissolved in water, and the solution made to 250 c.c.; 50 c.c. of this are taken, diluted with the same bulk of water, about 5 c.c. of pure concentrated sulphuric acid added, and, after warming to 60°, the permanganate solution from a burette is run in. At first the color does not disappear rapidly, but this soon alters, and as the liquid to be standardized is dropped in the color becomes instantly discharged as long as any of the salt remains unoxidized. As soon as the color becomes permanent, and the solution is of a very faint rose-color, the end point of the reaction is attained; 71 parts of ammonic oxalate are equivalent to 56 parts of iron.

Example: 1.620 grammes oxalate of ammonia was dissolved to 250 c.c., and 50 c.c., equal to .340 gramme of the salt, taken for the titration, which required 42 c.c. of the permanganate. Hence 42 c.c. is equivalent to .340 gramme oxalate, or

$$71 : 56 :: .340 : x = .2682.$$

$\frac{.2682}{42} = .006385$ gramme iron for 1 cubic centimeter of the standard solution. The standardizing should be done in duplicate.

Pure metallic iron in the form of pianoforte-wire may be used in the place of the ammonic oxalate, by the solution of a weighed portion of it in pure sulphuric acid in an atmosphere of carbonic acid or steam to prevent oxidation.*

ESTIMATION OF SOLUBLE MATTER.

To 25 grammes of the finely-powdered char are added 200 c.c. of warm water (not above 65°), and the mixture allowed to stand for a half-hour, with frequent agitation. The insoluble matter is allowed to settle, the clear liquid poured off through a filter, and about 100 c.c. more of water added to the residue, which is treated as before, but for a shorter time, and, after settling, the supernatant liquid is filtered. The washing is repeated once more, the undissolved residue, together with the liquid, is transferred to the filter, and the insoluble matter remaining on it is washed until free from anything soluble. The combined

* Chlorine is set free when permanganate is added to a solution containing hydrochloric acid, which tends to introduce an error in the results of iron determinations made under such conditions. For the small amounts of iron in bone-char, however, the influence of this error in the above estimation may be altogether neglected. (See Fresenius' *Quant. Analysis*, Am. ed., 198.)

filtrates are evaporated in a platinum dish over a water-bath to dryness, with the addition of sufficient hydrochloric acid to give the solution a faint acid reaction, in order to prevent the escape of ammonia as carbonate and sulphide. The addition of the acid alters the combination of some of the bodies present, but the error introduced is slight.

The weight of the dried residue is the *total soluble matter*. After the last weighing the dish is ignited only for a time sufficient to burn off the carbon, and the inorganic residue constitutes the *soluble mineral matter*, while the difference between this and the total is the *organic soluble matter.*

W. Thorn* determines the organic matter in char by taking 50 grammes, heating with 25 c.c. soda-lye of 1.4 sp. gr. and 200 c.c. of water, and washing out the yellow solution with hot water. The alkaline solution obtained is supersaturated with sulphuric acid and titred with solution of potassium permanganate; 5 parts of organic matter = 1 part of salt, or 1 c.c. of normal permanganate solution = .158 gramme organic matter. This process may give good comparative results.

ESTIMATION OF SUGAR.

One hundred grammes of powdered char are heated with two or three times its weight of hot water for a half-hour, with occasional shaking; the clear solution, after settling, is filtered and the washing repeated twice; and finally the residue is brought on the filter and further washed until all soluble matter is removed. The filtrates are evaporated on

* *Wagner's Jahresb.*, 1875, 812.

a water-bath, in a porcelain dish, to about 80 c.c., caustic alkali being added to very slight alkaline reaction. The solution, after cooling, is made up to 100 c.c., and polarized after the free alkali has been saturated by acetic acid. If the amount of sugar is too small for the saccharimetric test, resort must be had to the inversion method, which is more accurate in this case and should be generally preferred. When this method is used the liquid may be evaporated with the addition of hydrochloric acid to invert the sugar, and the invert-sugar formed estimated by Fehling's method, either gravimetrically or volumetrically. When the char is properly washed the amount of sugar remaining in it is extremely small, and cannot be estimated by the polarimetric method.

ESTIMATION OF SPECIFIC GRAVITY.

1. **Apparent Specific Gravity.**—This is simply a comparison of the weight of equal volumes of water and char. The determination is made by filling a tared half-litre flask with char, accompanied with a gentle shaking, and taking the weight, which, after subtracting that of the flask, gives that of the half-litre of char. This divided by the weight of the same volume of water gives the apparent specific gravity.

This determination is of little use in estimating the value of char, unless the size of the grains in each sample compared is the same, and also that all conditions of the experiments are similar, such as the amount of shaking, etc. The apparent specific gravity is often expressed in another form as the weight of one cubic foot of the material; it may be calculated by the formula—

$$P = \frac{W \times 28.315}{453.6},$$

in which P equals the avoirdupois pounds in one cubic foot, and W the weight in grammes of one litre.

2. **Absolute or Real Specific Gravity.**—Place 50 grammes of the char in a tared 100-c.c. flask partially filled with distilled water, boil for some minutes to free from air, fill to 100 c.c. after cooling, and weigh. The calculation is illustrated by an example:

Char + flask + water................	180	grammes.
Char (50 grammes), flask (55 grammes).	105	"
Water............................	75	"

As the flask without char would hold 100 grammes of water, 25 grammes must have been displaced by char; hence

$$\frac{50}{25} = 2.000 \text{ sp. gr.}$$

The sp. gr. thus obtained is independent of the pores in the coal, and hence the greater the density, other things being equal, the poorer the quality of the char.

ESTIMATION OF THE ABSORPTIVE POWER.

I. **The Absorptive Power for Color and Soluble Matter determined on the Large Scale** (by Stammer's colorimeter, Fig. 33).

For this purpose the sugar solution is compared before and after filtration. The color of the liquors referred to the sugar present, is determined according to directions given in chap. ix. ; the difference in the solutions, before and after filtration represents the color absorbed. The amount of organic matter and salts taken up is also determined. It is essential, in order that these results should be of any value, that an average sample of each liquor should be operated upon, and it should be especially assured that no "sweet water" or syrup foreign to the liquors under examination be allowed to mix with them. The following is an example taken from actual working:

	Liquor.		Per cent. absorbed.
	Before filtration.	After filtration.	
Sugar..................	88.40	91.30	....
Glucose..................	5.23	5.67	....
Organic matter not sugar.........	5.41	2.38	56.0
Ash..................	.96	.65	32.3
	100.00	100.00	
Color referred to per cent. of sugar.	54	23	57.4

An estimation made as the above, if from correct samples, is of great value in forming an opinion as to the condition of the char in actual use, and should never be neglected where it is practicable to make it.

II. **Estimation of the Decolorizing Power in the Laboratory.**—This method has to be resorted to in the examination of char for purchase, or when a comparatively small sample is at the disposal of the operator. Dilute any sample of dark-colored molasses or syrup with five times its weight of water, and determine the color of the solution by the colorimeter. Next weigh 100 grammes of the coal to be examined, place it in a flask with 300 c.c. of the dilute sugar-liquor, heat on a water-bath to 100° for one hour, shaking at intervals, filter, allow to cool, make up any loss by evaporation by adding water, and observe the color again. The difference before and after this treatment represents the decolorizing power of the char. Example :

Before filtration........................	28
After "	18
	10

$\frac{10}{28} = 35.70$ per cent. of the original color absorbed.

It is important that the conditions in all respects should be the same, in this determination, when made at different times and on different samples. A well-defined method of procedure should be laid down, not to be varied from in any case, as—the char should always be used of one degree of fineness, and the sieve used to bring it to that if necessary ; the dilution and composition of the sugar-liquor should be as nearly as possible the same, as well as the proportion by weight of char to volume of liquor employed, degree of heat, time of the experiment, amount of shaking, etc.

ESTIMATION OF THE COLOR WITH DUBOSCQ'S COLORIMETER.

A A′ are two glass cylinders with plane bottoms (Figure 43), one of which is destined to receive the solution to be examined, and the other the standard liquor. Two tubes, B B′, of small diameter, closed at the lower ends by glass plates, and capable of upward and downward motion by means of a rack and pinion, are placed behind the instrument on the upright support. Each pinion has a pointer, which measures upon the divided scale the respective distances between the bottoms A A′ and B B′. The upper part of the instrument carries a system of prisms and a small telescope, D, which enables the operator to see the relative color of the solutions under examination after the manner of Stammer's colorimeter. A movable mirror placed at E throws the light through the solutions. In the form of the

Fig. 43.

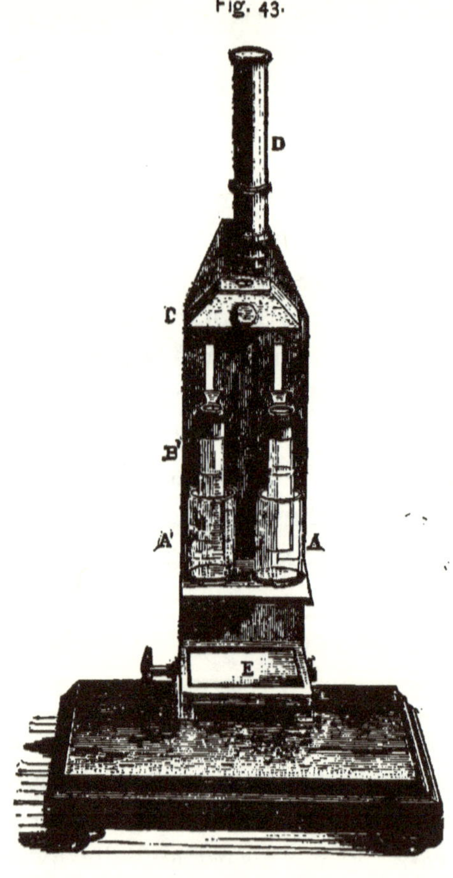

apparatus shown the light reflected from E had a tendency to enter the tubes out of the exact centre. To remedy this Duboscq has lately made an improvement which consists in interposing between the mirror and the bottom of the tubes a system of two birefrigerent prisms, joined together at their bases so that the line of contact is exactly between A A′ extended.

The type-liquor is made by dissolving two grammes of caramel in water and diluting the solution to one litre. The caramel is prepared by heating refined sugar for one and a half hours on a paraffin-bath at a temperature not above 215°. A little of the mass should be taken out at the end of that time, inverted by heating with acid, and tested with copper-liquor for sugar; if any is present the heating must be continued until all the sugar has been decomposed. The caramel should be preserved in a tight bottle.

The use of Duboscq's colorimeter is as follows: Of raw sugar or syrup a known weight is dissolved in water, and the solution made to 100 c.c. and observed in the instrument. At the commencement of the experiment B B′ should stand at the same height, A′ being filled with the type-liquor and A with the sugar solution to be examined. If the colors of the luminous field of the apparatus appear unequal on either side of the vertical line, A is elevated or depressed by its appropriate pinion until equality of tint is obtained. The relative heights of the two columns of colored solution is measured in millimetres on the back of the instrument. The proportion of caramel or coloring matter is in inverse ratio to the heights of the liquid columns. Thus the standard caramel solution contains in 100 c.c. .200 gramme caramel, from which datum the percentage of coloring matter in a sugar solution of known strength may be readily calculated. Example: The heights of the liquid columns as measured on the scale are 20 mm. for the standard and 40 mm. for the solution to be compared; then, as 100 c.c. of the former contain .200 gramme caramel, we have

$$20 : 40 :: x : .200 = .100 \text{ gramme}$$

coloring matter in 100 c.c. of the solution tested, which divided by the amount of the original substance in 100 c.c. gives the percentage.

To test the decolorizing power of char by Duboscq's Method, a weighed portion of the char is mixed with a known volume of the type-liquor and heated for a half or one hour, with the precautions as to the relative conditions of experiments mentioned on page 329. The difference in color before and after decolorizing shows, by a calculation similar to the one above, the actual amount of caramel removed.

Duboscq's process may be used with Stammer's instrument, and, in as far as it relates to the decolorizing power of char, Duboscq's is theoretically a better method than Stammer's, because the type-liquor is of supposed constant composition in the former, thus approaching an absolute standard, while with the latter a solution of constant color or composition cannot always be had. Unfortunately for the accuracy of Duboscq's process, it is practically exceedingly difficult, if not impossible, to prepare the standard caramel solution at different times having exactly the same tinctorial power. This fault in the method does not, however, affect the general usefulness of the colorimeter as a measurer of color in sugar solutions.

CORENWINDER'S METHOD FOR ESTIMATING THE ABSORBING POWER OF CHAR BY A SOLUTION OF CALCIC SUCRATE.

To prepare the sucrate solution, dissolve 125 grammes of

sugar in 600 to 800 c.c. of water, add 15 to 20 grammes caustic lime, boil five minutes, allow to cool, and make to one litre. To 100 c.c. of this solution 50 grammes of the char to be examined are added, and the whole left to stand for an hour, with frequent agitation, and then filtered. When the operation is finished, a part of the lime salt is absorbed, and this is to be estimated. By determining the amount of lime by standard nitric acid in 50 c.c. before and after the action of the char, the desired result is obtained. (See estimation of alkalinity, page 258.)

TEST TO DETERMINE THE COMPLETENESS OF WASHING AND BURNING.

This is by boiling for a few moments small portions of the char with solutions of sodium or potassium hydrate of 20° B. A yellow or brown color shows the presence of organic matter, and the greater the amount the greater the intensity of the color. The char, if properly burned, will give no reaction by this test, while the simply washed but unburned article should not give more than a lemon-color for ordinarily good syrups filtered, or a somewhat darker color for lower products.

One circumstance may render the indications of the above test fallacious—that is, when iron and sulphide of calcium are present in the char to a considerable extent, as often happens in old chars, they act upon each other in the presence of caustic alkali, producing a yellowish or greenish tint in the solution, due to the formation of ferrous sulphide. This indication may be distinguished from that of the simple action of alkali on organic matter by the tendency of the solution in the former case to acquire a tint of green on standing, and by the fact that the reaction readily takes place in the cold. Char capable of giving a decided green color under the above circumstances generally carries a high percentage of the sulphide, and is entirely unfit for the best uses (page 308).

APPENDIX.

NOTE

CONCERNING THE ACTION OF THE ORGANIC MATTER NOT SUGAR OCCURRING IN CANE AND BEET PRODUCTS, ON ALKALINE SOLUTION OF COPPER OXIDE.

It has been asserted that the organic matters present in impure commercial sugars and syrups have a considerable reductive effect on the copper solution employed in Fehling's method and its various modifications for the estimation of invert-sugar. To avoid this supposed source of error, it has accordingly been recommended that sugar solution before testing, should be treated with excess of basic lead acetate, filtered, the metallic salt remaining in solution precipitated with sulphurous acid, and the resulting liquid after filtering again, used for the sugar determination.

In order to prove whether the organic matters acted as asserted with Fehling's solution, the author has made some experiments on the most impure saccharine material obtainable from a variety of sources. The manner of conducting the experiments was as follows: The hot solution of the substance operated upon was treated with excess of basic lead acetate and the precipitate washed thoroughly with a large excess of hot water. To be assured that no sugar should remain in the precipitate as lead sucrate, the washed magma, after diffusion through water, was saturated with carbonic acid by allowing the gas to bubble through the diffused mass for six or eight hours, or longer. After this treatment the precipitate was again well washed, and then, after mixing with water, decomposed by sulphuretted hydrogen, the lead sulphide filtered off, and the resulting dissolved organic matters evaporated to dryness at a gentle heat and weighed. The substance thus separated was heated with Fehling's solution (Violette's), near the boiling-point, for some minutes, the resulting cuprous oxide converted in cupric oxide, and weighed, special correction being made for the filter-ash (page 203).

The results are given below; .100 gramme organic matter of the different origins given, caused the reduction of the following amounts of copper oxide:

West India molasses....	.027	gramme	CuO
Residual syrup from sugar-refining ..	.030	"	"
Beet-molasses......................	.0124	"	"
Muscovado raw sugar	.0246	"	"
Manilla "	.0170	"	"
.100 gramme invert-sugar reduces..	.2206	"	"

In order to test whether any sugar might have been retained in the precipitates before decomposition with lead, a control experiment was made by adding tartrate of soda and potash to a solution of pure invert-sugar, and precipitating with basic acetate of lead. This produced a voluminous precipitate similar to that thrown down from impure sugar solutions. The tartrate of lead was treated in the same manner as the compounds of lead and organic matter, as detailed above, and the tartaric acid obtained after the decomposition with sulphuretted hydrogen was heated with the copper solution.

.100 gramme reduced .002 gramme CuO.

From this it may be concluded that little or no sugar was retained in the organic lead compounds operated upon.

As a general result of these experiments, it may be proved by calculation that organic matters in impure sugars have too small an influence upon the results of the copper test to make it necessary to remove them from the sugar solutions in most cases. When the saccharine material contains a considerable proportion of these compounds, for very accurate work such removal may be desirable.

TABLES.

I.

PARTIAL LIST OF THE ATOMIC WEIGHTS.

	Symbol.	Old.	New.
Aluminium	Al	13.75	27.5
Antimony	Sb	122.	122.
Arsenic	As	75.	75.
Barium	Ba	68.5	137.
Bismuth	Bi	210.	210.
Boron	B	11.	11.
Bromine	Br	80.	80.
Calcium	Ca	20.	40.
Carbon	C	6.	12.
Chlorine	Cl	35.5	35.5
Chromium	Cr	26.2	52.4
Cobalt	Co	29.5	59.
Copper	Cu	31.7	63.4
Fluorine	F	19.	19.
Gold	Au	196.	196.
Hydrogen	H	**1.**	**1.**
Iodine	I	127.	127.
Iron	Fe	28.	56.
Lead	Pb	103.5	207.
Magnesium	Mg	12.	24.
Manganese	Mn	27.5	55.
Mercury	Hg	100.	200.
Nickel	Ni	29.5	59.
Nitrogen	N	14.	14.
Oxygen	O	**8.**	**16.**
Phosphorus	P	31.	31.
Platinum	Pt	98.94	197.9
Potassium	K	39.1	39.1
Silicon	Si	14.	28.
Silver	Ag	108.	108.
Sodium	Na	23.	23.
Strontium	Sr	43.75	87.5
Sulphur	S	16.	32.
Tin	Sn	59.	118.
Zinc	Zn	32.5	65.

II.

Density of Sugar Solutions of various Temperatures and Concentrations (after Gerlach).

Temp. C.	0 per ct.	5 per ct.	10 per ct.	15 per ct.	20 per ct.	25 per ct.	30 per ct.	35 per ct.	40 per ct.	45 per ct.	50 per ct.	55 per ct.	60 per ct.	65 per ct.	70 per ct.	75 per ct.
0	1.0007	1.0210	1.0418	1.0636	1.0861	1.1094	1.1327	1.1566	1.1845	1.2113	1.2390	1.2676	1.2972	1.3276	1.3590	1.3916
5	1.0009	1.0210	1.0417	1.0632	1.0856	1.1089	1.1329	1.1577	1.1833	1.2098	1.2372	1.2657	1.2951	1.3255	1.3568	1.3892
10	1.0008	1.0203	1.0413	1.0627	1.0849	1.1079	1.1318	1.1565	1.1819	1.2082	1.2355	1.2638	1.2931	1.3233	1.3545	1.3869
15	1.0003	1.0202	1.0406	1.0619	1.0839	1.1067	1.1305	1.1550	1.1802	1.2065	1.2337	1.2619	1.2910	1.3211	1.3522	1.3845
17.5	1.0000	1.0197	1.0401	1.0613	1.0832	1.1060	1.1296	1.1540	1.1794	1.2056	1.2327	1.2609	1.2900	1.3200	1.3512	1.3833
20	0.9956	1.0191	1.0395	1.0606	1.0825	1.1052	1.1288	1.1532	1.1785	1.2046	1.2317	1.2599	1.2889	1.3189	1.3500	1.3822
25	0.9666	1.0179	1.0382	1.0592	1.0810	1.1036	1.1271	1.1514	1.1766	1.2026	1.2296	1.2578	1.2867	1.3166	1.3477	1.3798
30	0.9973	1.0166	1.0367	1.0576	1.0794	1.1018	1.1253	1.1495	1.1746	1.2005	1.2274	1.2555	1.2844	1.3142	1.3454	1.3773
35	0.9958	1.0150	1.0351	1.0560	1.0775	1.1000	1.1233	1.1474	1.1726	1.1982	1.2251	1.2531	1.2820	1.3117	1.3430	1.3748
40	0.9942	1.0133	1.0332	1.0540	1.0756	1.0978	1.1212	1.1452	1.1701	1.1958	1.2227	1.2507	1.2794	1.3091	1.3405	1.3722
45	0.9923	1.0113	1.0311	1.0519	1.0733	1.0956	1.1183	1.1423	1.1677	1.1933	1.2202	1.2481	1.2768	1.3065	1.3380	1.3695
50	0.9903	1.0052	1.0290	1.0496	1.0710	1.0932	1.1164	1.1403	1.1651	1.1907	1.2175	1.2454	1.2741	1.3038	1.3352	1.3667
55	0.9881	1.0069	1.0266	1.0472	1.0685	1.0907	1.1138	1.1378	1.1624	1.1879	1.2148	1.2426	1.2712	1.3010	1.3325	1.3639
60	0.9857	1.0044	1.0242	1.0448	1.0660	1.0881	1.1114	1.1351	1.1594	1.1851	1.2119	1.2397	1.2683	1.2982	1.3295	1.3610
65	0.9831	1.0018	1.0217	1.0421	1.0634	1.0854	1.1084	1.1324	1.1567	1.1822	1.2091	1.2367	1.2654	1.2953	1.3265	1.3580
70	0.9804	0.9992	1.0190	1.0393	1.0607	1.0827	1.1056	1.1295	1.1538	1.1792	1.2062	1.2337	1.2624	1.2923	1.3233	1.3552
75	0.9775	0.9964	1.0162	1.0363	1.0578	1.0798	1.1027	1.1265	1.1509	1.1761	1.2031	1.2307	1.2593	1.2893	1.3201	1.3520
80	0.9745	0.9935	1.0132	1.0336	1.0547	1.0763	1.0997	1.1234	1.1479	1.1729	1.2000	1.2276	1.2562	1.2862	1.3168	1.3488
85	0.9714	0.9905	1.0101	1.0305	1.0515	1.0736	1.0965	1.1202	1.1448	1.1696	1.1967	1.2246	1.2529	1.2829	1.3136	1.3456
90	0.9683	0.9872	1.0068	1.0272	1.0481	1.0701	1.0931	1.1169	1.1415	1.1664	1.1934	1.2213	1.2496	1.2796	1.3103	1.3423
95	0.9652	0.9838	1.0034	1.0237	1.0447	1.0663	1.0897	1.1135	1.1381	1.1630	1.1899	1.2179	1.2462	1.2763	1.3067	1.3390
100	0.9621	0.9804	0.9959	1.0202	1.0413	1.0634	1.0863	1.1099	1.1346	1.1597	1.1864	1.2145	1.2427	1.2728	1.3032	1.3356

III.

TABLE SHOWING THE RELATION OF TEMPERATURE, DENSITY, AND DEGREES BAUMÉ of SUGAR SOLUTIONS (FLOURENS).

Temp. C.	Per cent. of sugar.	Baumé,		Density,	
		At obser. temp.	At 15° C.	At obser. temp.	At 15° C.
0	64.70	35.30	34.60	1.3235	1.3150
5	65.00	35.35	34.90	1.3243	1.3190
10	65.50	35.45	35.00	1.3255	1.3225
15	66.00	35.50	35.50	1.3260	1.3260
20	66.50	35.60	35.75	1.3275	1.3290
25	67.20	35.80	36.25	1.3300	1.3355
30	68.00	36.00	36.70	1.3325	1.3405
35	68.80	36.20	37.10	1.3350	1.3460
40	69.75	36.40	37.50	1.3375	1.3510
45	70.80	36.75	38.10	1.3410	1.3590
50	71.80	37.10	38.70	1.3460	1.3660
55	72.80	37.50	39.30	1.3510	1.3740
60	74.00	37.90	39.90	1.3560	1.3820
65	75.00	38.30	40.55	1.3615	1.3910
70	76.10	38.60	41.10	1.3650	1.3980
75	77.20	39.00	41.70	1.3700	1.4060
80	78.35	39.30	42.20	1.3740	1.4130
85	79.50	39.65	42.80	1.3790	1.4220
90	80.60	39.95	43.30	1.3820	1.4290
95	81.60	40.10	43.70	1.3850	1.4300
100	82.50	40.30	44.10	1.3875	1.4400

IV.

TABLE SHOWING THE RELATION OF DENSITY, DEGREES BAUMÉ, AND BOILING-POINTS OF SUGAR SOLUTIONS (FLOURENS).

Boiling temp. C.	Baumé,		Density,		Per cent. of sugar.
	At observed temp.	At 15° C.	At observed temp.	At 15° C.	
104.5	32.20	36.25	1.2872	1.3350	67.25
105.	33.20	37.25	1.2990	1.3480	69.1
105.5	34.20	38.30	1.3106	1.3613	71.2
106.	35.00	39.10	1.3200	1.3720	72.4
106.5	35.50	39.65	1.3260	1.3780	73.4
107.	36.00	40.15	1.3325	1.3855	74.4
107.5	36.50	40.70	1.3385	1.3925	75.2
108.	37.00	41.10	1.3450	1.3985	76.4
108.5	37.50	41.75	1.3510	1.4080	77.4
109.	37.90	42.10	1.3562	1.4120	77.8
109.5	38.25	42.50	1.3606	1.4180	78.7
110.	38.50	42.80	1.3640	1.4215	79.5
110.5	38.75	43.00	1.3670	1.4245	80.0
111.	39.00	43.30	1.3700	1.4290	80.6
111.5	39.30	43.65	1.3740	1.4335	81.4
112.	39.60	44.00	1.3770	1.4380	82.2
112.5	39.80	44.20	1.3810	1.4415	82.9
113.	40.00	44.40	1.3835	1.4500	83.6
114.	40.30	....	1.3875	....	84.2
115.	40.60	....	1.3915	....	85.2
116.	40.90	....	1.3955	....	85.8
117.	41.20	....	1.4000	....	86.5
118.	41.45	....	1.4030	....	87.2
119.	41.65	....	1.4060	....	87.9
120.	41.90	....	1.4085	....	88.5
125.	42.80	....	1.4215	....	91.2
130.	43.50	....	1.4315	....	92.2

V.

Boiling-points of Sugar Solutions (after Gerlach).

Per cent. sugar.	Boiling-point, C.
10.	100.4
20.	100.6
30.	101.0
40.	101.5
50.	102.0
60.	103.0
70.	106.5
79.	112.0
90.8	130.0

VI.

Volumes of Sugar Solutions at Different Temperatures (Gerlach).

Temp. C.	10 per cent.	20 per cent.	30 per cent.	40 per cent.	50 per cent.
0	10000	10000	10000	10000	10000
5	10004.5	10007	10009	10012	10016
10	10012	10016	10021	10026	10032
15	10021	10028	10034	10042	10050
20	10033	10041	10049	10058	10069
25	10048	10057	10066	10075	10088
30	10064	10074	10084	10094	10110
35	10082	10092	10103	10114	10132
40	10101	10112	10124	10136	10156
45	10122	10134	10146	10160	10180
50	10145	10156	10170	10184	10204
55	10170	10183	10196	10210	10229
60	10197	10209	10222	10235	10253
65	10225	10236	10249	10261	10278
70	10255	10265	10277	10287	10306
75	10284	10295	10306	10316	10332
80	10316	10325	10335	10345	10360
85	10347	10355	10365	10375	10388
90	10379	10387	10395	10405	10417
95	10411	10418	10425	10435	10445
100	10442	10450	10456	10465	10457

VII.

THE AMOUNT OF LIME CONTAINED IN MILK OF LIME OF VARIOUS DENSITIES (MATEGCZEK).

Degree Baumé.	Degree Brix.	1 K. CaO contained in — litres milk of lime.	Degree Baumé.	Degree Brix.	1 K. CaO contained in — litres milk of lime.
10.	18.0	7.50	21.	38.3	4.28
11.	20.0	7.10	22.	40.2	4.16
12.	21.7	6.70	23.	42.0	4.05
13.	23.5	6.30	24.	43.9	3.95
14.	25.3	5.88	25.	45.8	3.87
15.	27.2	5.50	26.	47.7	3.81
16.	29.	5.25	27.	49.6	3.75
17.	30.9	5.01	28.	51.6	3.70
18.	32.7	4.80	29.	53.5	3.65
19.	34.6	4.68	30.	55.5	3.60
20.	36.5	4.42			

VIII.

DENSITY OF LIME SUCRATE SOLUTIONS (PELIGOT).

Per cent. of sugar.	Density of sugar solutions.	Density when saturated with CaO.	The sucrate solution contains in 100 parts:	
			CaO	Sugar.
40.0	1.122	1.179	21.0	79.0
37.5	1.116	1.175	20.8	79.2
35.0	1.110	1.166	20.5	79.5
32.5	1.103	1.159	20.3	79.7
30.0	1.096	1.148	20.1	79.9
27.5	1.089	1.139	19.9	80.1
25.0	1.082	1.128	19.8	80.2
22.5	1.075	1.116	19.3	80.7
20.0	1.068	1.104	18.8	81.2
17.5	1.060	1.092	18.7	81.3
15.0	1.052	1.080	18.5	81.5
12.5	1.044	1.067	18.3	81.7
10.0	1.036	1.053	18.1	81.9
5.0	1.027	1.040	16.9	83.1
2.5	1.018	1.026	15.3	84.7

IX.

TABLE SHOWING THE WEIGHT, AND SOLID MATTER CONTAINED IN THE CUBIC FOOT AND U. S. GALLON OF SUGAR SOLUTIONS, TOGETHER WITH THE CORRESPONDING DENSITIES, PERCENTAGES OF SOLID MATTER, AND DEGREES BAUMÉ.

Degree Baumé.	Deg. Brix. Per cent. of solid matter.	Weight of 1 cu. ft. lbs.	Solid matter in 1 cu. ft. lbs.	Weight of 1 gal. lbs.	Solid matt'r in 1 gal. lbs.	Specific Gravity.	Degree Beaumé.	Deg. Brix. Per cent. of solid matter.	Weight of 1 cu. ft. lbs.	Solid matt'r in 1 cu. ft. lbs.	Weight of 1 gal. lbs.	Solid matt'r in 1 gal. lbs.	Specific Gravity.
.5	.9	62.60	.56	8.37	.07	1.0035	23.0	42.1	74.20	31.23	9.92	4.16	1.1903
1.0	1.8	62.83	1.13	8.40	.15	1.0070	23.5	43.0	74.50	32.03	9.96	4.28	1.1950
1.5	2.7	63.05	1.70	8.43	.23	1.0105	24.0	44.0	74.87	32.94	10.01	4.40	1.2003
2.0	3.6	63.27	2.24	8.46	.30	1.0141	24.5	44.9	75.17	33.75	10.05	4.51	1.2051
2.5	4.5	63.50	2.86	8.49	.38	1.0177	25.0	45.9	75.47	34.64	10.09	4.62	1.2104
3.0	5.5	63.73	3.50	8.52	.47	1.0217	25.5	46.8	75.77	35.46	10.13	4.74	1.2153
3.5	6.3	63.95	4.04	8.55	.54	1.0249	26.0	47.7	76.07	36.28	10.17	4.85	1.2202
4.0	7.2	64.17	4.63	8.58	.62	1.0286	26.5	48.7	76.44	37.23	10.22	4.97	1.2256
4.5	8.1	64.40	5.21	8.61	.70	1.0323	27.0	49.6	76.74	38.06	10.26	5.09	1.2306
5.0	9.0	64.62	5.83	8.64	.78	1.0360	27.5	50.6	77.11	39.02	10.31	5.22	1.2361
5.5	9.9	64.85	6.43	8.67	.86	1.0397	28.0	51.5	77.42	39.87	10.35	5.33	1.2411
6.0	10.8	65.07	7.03	8.70	.94	1.0435	28.5	52.5	77.72	40.80	10.39	5.45	1.2467
6.5	11.7	65.30	7.63	8.73	1.02	1.0472	29.0	53.5	78.09	41.78	10.44	5.58	1.2523
7.0	12.6	65.52	8.24	8.76	1.10	1.0510	29.5	54.4	78.39	42.64	10.48	5.69	1.2574
7.5	13.5	65.82	8.90	8.80	1.19	1.0549	30.0	55.4	78.76	43.63	10.53	5.83	1.2631
8.0	14.4	66.05	9.50	8.83	1.27	1.0587	30.5	56.4	79.06	44.59	10.57	5.95	1.2689
8.5	15.4	66.27	10.17	8.86	1.36	1.0630	31.0	57.3	79.44	45.52	10.62	6.09	1.2741
9.0	16.3	66.50	10.84	8.89	1.45	1.0670	31.5	58.3	79.81	46.53	10.67	6.22	1.2799
9.5	17.1	66.72	11.41	8.92	1.53	1.0704	32.0	59.3	80.18	47.55	10.72	6.36	1.2859
10.0	18.0	67.02	12.06	8.96	1.62	1.0744	32.5	60.3	80.56	48.50	10.77	6.49	1.2911
10.5	18.9	67.24	12.71	8.99	1.70	1.0784	33.0	61.2	80.93	49.53	10.82	6.62	1.2970
11.0	19.9	67.45	13.37	9.02	1.79	1.0829	33.5	62.2	81.31	50.56	10.87	6.76	1.3030
11.5	20.8	67.77	14.10	9.06	1.88	1.0869	34.0	63.1	81.69	51.55	10.92	6.90	1.3084
12.0	21.7	68.07	14.77	9.10	1.97	1.0909	34.5	64.1	81.98	52.55	10.96	7.03	1.3154
12.5	22.6	68.29	15.42	9.13	2.06	1.0950	35.0	65.2	82.43	53.74	11.02	7.18	1.3211
13.0	23.5	68.52	16.10	9.16	2.15	1.0991	35.5	66.2	82.80	54.81	11.07	7.33	1.3272
13.5	24.4	68.81	16.79	9.20	2.25	1.1033	36.0	67.2	83.16	55.88	11.12	7.47	1.3333
14.0	25.3	69.04	17.47	9.23	2.34	1.1074	36.5	68.2	83.54	56.97	11.17	7.62	1.3395
14.5	26.2	69.34	18.17	9.27	2.42	1.1116	37.0	69.2	83.93	58.08	11.22	7.76	1.3458
15.0	27.2	69.64	18.94	9.31	2.53	1.1163	37.5	70.2	84.29	59.17	11.27	7.91	1.3521
15.5	28.1	69.86	19.63	9.34	2.62	1.1206	38.0	71.2	84.74	60.33	11.33	8.07	1.3585
16.0	29.0	70.16	20.34	9.38	2.72	1.1249	38.5	72.2	85.12	61.46	11.38	8.21	1.3649
16.5	29.9	70.39	21.05	9.41	2.81	1.1292	39.0	73.2	85.56	62.63	11.44	8.37	1.3714
17.0	30.8	70.68	21.77	9.45	2.91	1.1335	39.5	74.2	85.94	63.77	11.49	8.52	1.3780
17.5	31.8	70.99	22.57	9.49	3.01	1.1384	40.0	75.3	86.24	64.94	11.55	8.67	1.3846
18.0	32.7	71.28	23.30	9.53	3.11	1.1427	40.5	76.3	86.76	66.19	11.60	8.85	1.3913
18.5	33.6	71.51	24.03	9.56	3.21	1.1472	41.0	77.3	87.21	67.41	11.66	9.01	1.3981
19.0	34.6	71.88	24.87	9.61	3.32	1.1521	41.5	78.3	87.65	68.63	11.72	9.18	1.4049
19.5	35.5	72.11	25.60	9.64	3.42	1.1565	42.0	79.4	88.05	69.91	11.77	9.34	1.4118
20.0	36.4	72.40	26.35	9.68	3.51	1.1611	42.5	80.4	88.49	71.15	11.83	9.51	1.4187
20.5	37.4	72.70	27.19	9.72	3.62	1.1662	43.0	81.5	88.99	72.48	11.90	9.69	1.4267
21.0	38.3	73.00	27.96	9.76	3.74	1.1707	43.5	82.5	89.39	73.75	11.95	9.86	1.4328
21.5	39.2	73.30	28.73	9.80	3.84	1.1753	44.0	83.6	89.83	75.10	12.01	10.04	1.4400
22.0	40.1	73.60	29.51	9.84	3.95	1.1799	44.5	84.6	90.23	76.37	12.07	10.21	1.4472
22.5	41.1	73.90	30.37	9.88	4.07	1.1851	45.0	85.7	90.73	77.75	12.13	10.39	1.4545

X.

TABLE SHOWING THE EQUIVALENCE OF THE **Centigrade** THERMOMETER SCALE WITH THAT OF **Fahrenheit.**

C.	F.	C.	F.	C.	F.
+100	+212.	+63	+145.4	+26	+78.8
99	210.2	62	143.6	25	77.
98	208.4	61	141.8	24	75.2
97	206.6	60	140.	23	73.4
96	204.8	59	138.2	22	71.6
95	203.	58	136.4	21	69.8
94	201.2	57	134.6	20	68.
93	199.4	56	132.8	19	66.2
92	197.6	55	131.	18	64.4
91	195.8	54	129.2	17	62.6
90	194.	53	127.4	16	60.8
89	192.2	52	125.6	15	59.
88	190.4	51	123.8	14	57.2
87	188.8	50	122.	13	55.4
86	186.6	49	120.2	12	53.6
85	185.	48	118.4	11	51.8
84	183.2	47	116.6	10	50.
83	181.4	46	114.8	9	48.2
82	179.6	45	113.	8	46.4
81	177.8	44	111.2	7	44.6
80	176.	43	109.4	6	42.8
79	174.2	42	107.6	5	42.
78	172.4	41	105.8	4	39.2
77	170.6	40	104.	3	37.4
76	168.8	39	102.2	2	35.6
75	167.	38	100.4	1	33.8
74	165.2	37	98.6	0	32.
73	163.4	36	96.8	−1	30.2
72	161.6	35	95.	2	28.4
71	159.8	34	93.2	3	26.6
70	158.	33	91.4	4	24.8
69	156.2	32	89.6	5	23.
68	154.4	31	87.8	6	21.2
67	152.6	30	86.	7	19.4
66	150.8	29	84.2	8	17.6
65	149.	28	82.4	9	15.8
64	147.2	27	80.6	10	14.

XI.

TABLE SHOWING THE EQUIVALENCE OF THE **Fahrenheit** THERMOMETER SCALE WITH THAT OF THE **Centigrade.**

F.	C.	F.	C.	F.	C.
+212	+100.	+170	+76.67	+128	+53.33
211	99.44	169	76.11	127	52.78
210	98.89	168	75.55	126	52.22
209	98.33	167	75.	125	51.67
208	97.78	166	74.44	124	51.11
207	97.22	165	73.89	123	50.55
206	96.67	164	72.33	122	50.
205	96.11	163	72.78	121	49.44
204	95.55	162	71.22	120	48.89
203	95.	161	71.67	119	48.33
202	94.44	160	71.11	118	47.78
201	93.89	159	70.55	117	47.22
200	93.33	158	70.	116	46.67
199	92.78	157	69.44	115	46.11
198	92.22	156	68.89	114	45.55
197	91.67	155	68.33	113	45.
196	91.11	154	67.78	112	44.44
195	90.55	153	67.22	111	43.89
194	90.	152	66.67	110	43.33
193	89.44	151	66.11	109	42.78
192	88.89	150	65.55	108	42.22
191	88.33	149	65.	107	41.67
190	87.78	148	64.44	106	41.11
189	87.22	147	63.89	105	40.55
188	86.67	146	63.33	104	40.
187	86.11	145	62.78	103	39.44
186	85.55	144	62.22	102	38.89
185	85.	143	61.67	101	38.33
184	84.44	142	61.11	100	37.78
183	83.89	141	60.55	99	37.22
182	83.33	140	60.	98	36.67
181	82.78	139	59.44	97	36.11
180	82.22	138	58.89	96	35.55
179	81.67	137	58.33	95	35.
178	81.11	136	57.78	94	34.44
177	80.55	135	57.22	93	33.89
176	80.	134	56.67	92	33.33
175	79.44	133	56.11	91	32.78
174	78.89	132	55.55	90	32.22
173	78.33	131	55.	89	31.67
172	77.78	130	54.44	88	31.11
171	77.22	129	53.89	87	30.55

XI.—(*Continued.*)

F.	C.	F.	C.	F.	C.
+86	+30.	+61	+16.11	+37	+2.78
85	29.44	60	15.55	36	2.22
84	28.89	59	15.	35	1.67
83	28.33	58	14.44	34	1.11
82	27.78	57	13.89	33	0.55
81	27.22	56	13.33	32	0.
80	26.67	55	12.78	31	0.55
79	26.11	54	12.22	30	−1.11
78	25.55	53	11.67	29	1.67
77	25.	52	11.11	28	2.22
76	24.44	51	10.55	27	2.78
75	23.89	50	10.	26	3.33
74	23.33	49	9.44	25	3.89
73	22.78	48	8.89	24	4.44
72	22.22	47	8.33	23	5.
71	21.67	46	7.78	22	5.55
70	21.11	45	7.22	21	6.11
69	20.55	44	6.67	20	6.67
68	20.	43	6.11	19	7.22
67	19.44	42	5.55	18	7.78
66	18.89	41	5.	17	8.33
65	18.33	40	4.44	16	8.89
64	17.78	39	3.89	15	9.44
63	17.22	38	3.33	14	10.
62	16.67				

XII.

CHANGES IN VOLUME OF SUGAR SOLUTIONS WHEN DILUTED WITH WATER (AFTER GERLACH).

Per cent of sugar in solution.	Found density at 17½° C.	Mean calculated density.	Volume after mixing.
70	1.3507	1.3507	1.0000
60	1.2900	1.28626	.99710
50	1.2329	1.22769	.99577
40	1.1794	1.17422	.99560
30	1.1295	1.12521	.99620
20	1.0832	1.08013	.99716
10	1.0404	1.03852	.99819
0	1.0000	1.0000	1.0000

INDEX.

A

B

www.ingramcontent.com/pod-product-compliance
Lightning Source LLC
Chambersburg PA
CBHW030255240426
43673CB00040B/978

* 9 7 8 3 3 3 7 4 1 4 3 2 0 *